Technical Engineering Writing

emerald PUBLISHING

ice Publishing

Technical Engineering Writing

A Practical Guide for Professional Engineers

Jean Billingsley
MBA CEng FIMechE CEnv CMgr FCMI

Published by Emerald Publishing Limited,
Floor 5, Northspring, 21-23 Wellington Street, Leeds LS1 4DL.

ICE Publishing is an imprint of Emerald Publishing Limited

Other ICE Publishing titles:
Successful Professional Reviews for Civil Engineers, Fifth edition
Patrick Waterhouse. 978-0-7277-6609-0
Communication for Professional Engineers, Second edition
Bill Scott and Bertil Billing. 978-0-7277-2630-8

A catalogue record for this book is available from the British Library

ISBN 978-1-83662-927-6

Cover photo: Shutterstock/Pixels Hunter

Commissioning Editor: Michael Fenton
Content Development Editor: Ryan Molyneux
Production Editor: Emma Sudderick
Project Manager: Rajachitra S

Typeset by KnowledgeWorks Global Limited

Printed and bound by CPI Group (UK) Ltd, Croydon, CR0 4YY

Contents

About the author

Jean Billingsley

Jean Billingsley is a Chartered Engineer and has over 35 years' experience in engineering and project management, holding senior management positions at both Rolls-Royce and Alstom, and has delivered multi-million pound projects. Jean moved into training and consulting in 2013 and has delivered training to many companies and organisations. She prepares and delivers training courses, specialising in written communication skills for engineers.

Abbreviations

5Ws and 1H	what, why, where, when, who and how
ABC	accuracy, brevity, clarity
AI	artificial intelligence
ANPR	automatic number plate recognition
APA	American Psychological Association
ASD	Aerospace, Security and Defence Industries Association of Europe
BIM	building information modelling
BLUF	bottom line up front
BOM	bill of materials
BSI	British Standards Institution
CAD	computer-aided design
CEFR	Common European Framework of Reference for Languages
CEng	Chartered Engineer
CPD	continuing professional development
CV	curriculum vitae
DSE	display screen equipment
EngTech	Engineering Technician
FANBOYS	for, and, nor, but, or, yet, so
GDPR	General Data Protection Regulation
GD&T	geometric dimensioning and tolerancing
GPT	Generative Pre-trained Transformer
HSE	Health and Safety Executive
ICE	Institution of Civil Engineers
IEng	Incorporated Engineer
IET	Institution of Engineering and Technology
IMechE	Institution of Mechanical Engineers
IP	intellectual property
IPR	intellectual property rights
IStructE	Institution of Structural Engineers
ITP	inspection & test procedure
ITT	invitation to tender
JARS	Journal Article Reporting Standards
KFD	know, feel, do
LASER	light amplification by stimulated emission of radiation
LLM	large language model
MLA	Modern Language Association of America
NDA	non-disclosure agreement
NLP	natural language processing
OED	*Oxford English Dictionary*

OUP	Oxford University Press
PAT	portable appliance testing
PDCA	plan, do, check, act
PM	project management
PPE	personal protective equipment
PRIDE	purpose, reader, integration, draft, edit
QMS	quality management system
R&D	research and development
RFP	request for proposal
RFQ	request for quotation
RICS	Royal Institution of Chartered Surveyors
RMC	ready-mixed concrete; residents' management company
SCUBA	self-contained underwater breathing apparatus
SI	International System of Units
SME	small and medium-sized enterprises; subject-matter expert
SOP	standard operating procedure
SOW	statement of work
SPaG	spelling, punctuation and grammar
SQEP	suitably qualified and experienced person
UK-SPEC	UK Standard for Professional Engineering Competence and Commitment
WCAG	Web Content Accessibility Guidelines
WIIFM	What's in it for me?

Jean Billingsley
ISBN 978-1-83662-927-6
https://doi.org/10.1108/978-1-83662-924-520261001

Chapter 1
Introduction

> Many new engineering graduates still lack the level of writing skills expected by their employers [...] Another contribution to the problem of poor writing skills arises from students not having a clear idea of what 'good' writing looks like.
>
> (Selwyn and Renaud-Assemat, 2020)

Most professional engineers do not go into engineering because they have a burning desire to write. Yet writing is an essential part of many engineers' work. Whether it is reporting progress, specifying requirements or making a proposal, engineers need to communicate technical information. It takes skill to be able to synthesise complex ideas and explain them to someone else, particularly if the reader is not a subject-matter expert.

> When asked whether their background education in the engineering field adequately prepared them for writing on the job, only 30 (50%) were confident of their writing abilities following their engineering degrees. The other 30 (50%), although they found their skills in technical writing to be useful, were not confident of its adequacy in the workplace.
>
> (Khan, 2019)

Although writing in industry is similar in some ways to writing while studying, there are differences. For a qualification, writing is to demonstrate learning to a knowledgeable reader. The purpose and the reader are quite different, and writing to a minimum word count can become a bad habit. Engineers need a benchmark of what good writing looks like.

This book is for early career professional engineers developing their writing skills. It is also for established professional engineers, partly as a refresher and as a reference guide for good practice in technical engineering writing. It is for civil engineers, mechanical engineers and any of the many disciplines of professional engineering. It should be useful for other scientific and technical disciplines too, although the examples used may be more relevant to engineers.

Communication is recognised as a core competence for anyone applying to be a Chartered Engineer, Incorporated Engineer or Engineering Technician. The UK Engineering Council licenses 39 professional engineering institutions, including the Institution of Civil Engineers (ICE), the Institution of Engineering and Technology (IET) and the Institution of Mechanical Engineers (IMechE). The Engineering Council also has other professional affiliates, including the Royal Institution of Chartered Surveyors (RICS).

Developing efficient and effective writing skills can help an engineering role feel easier and be more enjoyable. Writing long documents can seem daunting, particularly in the second quarter of the twenty-first century, with everyday electronic communications, social media and instant messaging. Engineers need to be able to explain things clearly. If a reader has to ask, 'So what?' then a document has failed in its purpose.

Communication is usually better when more than one form is used, for example, explaining the highlights of a design report in a design review or presenting a proposal in a meeting. Two-way communication helps to avoid misunderstanding, as there is an opportunity for questions and clarifications. However, today's workforce often operates remotely, perhaps hybrid, even globally. An engineer's writing, therefore, needs to be able to stand alone without further explanation.

Good writing skills can help engineers develop and progress in their careers, whatever their career goals. Written documents are the visible presentation of someone's work. Formal documents are usually retained in some form of document management system and may be viewed many years later. Creating a positive impression may also help raise an engineer's personal profile and likelihood of promotion.

While there are many books on English grammar and writing, there are few specifically for engineers, and they do not have much on planning and the writing process. This book aims to fill these gaps. It is written for engineers, understanding their challenges and typical experiences, and using their language and examples.

Since 2013, I have trained thousands of engineers in technical writing. I am proud to be a Chartered Engineer and a Fellow of IMechE. First and foremost, I am an engineer and my objective is to help other engineers with a practical guide. I continue to learn about writing myself, even in researching and developing this book. I hope to address common issues that I have seen first-hand and to provide advice that I know many engineers have found useful.

This book is written specifically for engineers to help them with the writing they do as part of their engineering work. It covers the general principles of writing, such as planning, the writing process, style and presentation. It will also explore the key points, typical structure and contents for some common engineering documents. The focus will be on technical writing for formal documents, such as technical reports.

The contents of this book are designed to develop logically and ideally should be read in full and in order for the best understanding. However, there is an index and many sections are cross-referenced so that the book can be used as a reference guide. It also includes several checklists to support writing.

This book is not about learning English, style, or spelling, punctuation and grammar (SPaG). There are many excellent works covering these, some of which are referenced. This book is focused on helping professional engineers to write more clearly and avoid ambiguity. While considering an international readership, it focuses on the British context and UK English, highlighting some of the differences from US English.

Nor is it a guide to using Microsoft Word or any other software. As this book is intended as a practical guide, it does cover some useful functions of Word, particularly hints and tips for saving time and improving presentation. There are also other word processors and software tools that can help with writing. Artificial intelligence is a powerful emerging tool, though to be used with care.

> Even with ChatGPT writing for you, being able to write yourself is the most important skill of the next decade.
>
> (Bartlett, 2026)

Writing well is a core skill that can help professional engineers perform better in their roles and progress in their careers. Like many aspects of engineering, having a standard for best practice and a process to follow with supporting checklists goes a long way. Writing is a skill that engineers can learn and add to their toolboxes.

REFERENCES

Bartlett S (2026) https://www.linkedin.com/posts/stevenbartlett-123_unpopular-opinion-writing-is-going-to-be-activity-7422322603048128512-Mn3v/ (accessed 29/01/2026).

Khan A (2019) The engineers' guide to technical writing: insights for budding engineers. *Advances in Language and Literary Studies* **10(4)**: 80–87.

Selwyn R and Renaud-Assemat I (2020) Developing technical report writing skills in first and second year engineering students: a case study using self-reflection. *Higher Education Pedagogies* **5(1)**: 19–29, 10.1080/23752696.2019.1710550.

Technical Engineering Writing: A Practical Guide for Professional Engineers
Jean Billingsley
ISBN 978-1-83662-927-6
https://doi.org/10.1108/978-1-83662-924-520261003

Chapter 2
Planning

Almost everything we do starts with planning. If we want to cook dinner, we start by deciding what we want to eat and buying the ingredients. If we only open the fridge at dinner time, we may be looking at empty shelves and end up with a rather disappointing meal. Engineers are very familiar with working with processes in many aspects of their work. Writing is no different – it is a process and starts with a plan.

Processes are often shown as a cycle, such as with project management (PM). Although a project is, by definition, temporary and unique, the PM process is often shown as a cycle, which includes lessons learnt leading into the next project. Similarly, effective writing comes from following a simple cycle, and engineers will improve with experience and supporting systems.

Engineers may be familiar with the Deming 'plan, do, check, act' (PDCA) cycle for continuous improvement and quality management. As for this tried-and-tested PDCA process, we can apply a similar approach to writing. Like any good process, it starts with planning and continues with draft, edit and review (Figure 2.1).

Figure 2.1 The writing process

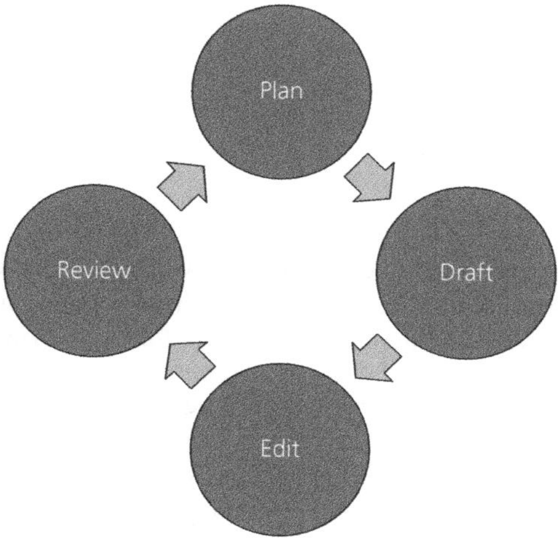

Some of the most important aspects of effective writing can be summarised in the acronym PRIDE (Billingsley, 2017):

- P – purpose
- R – reader
- I – integration
- D – draft
- E – edit.

We will explore the first three of these – purpose, reader and integration – in this chapter on planning. The next chapter, on writing, covers the draft and edit stages and includes the review element of this four-stage process.

2.1. Rationale for engineering documents

Professional engineers have to prepare a wide variety of different types of engineering and technical documentation and deliverables. There are very few engineering activities that involve little or no writing, such as project plans, technical drawings or schematics, computer-aided design (CAD) models, a bill of materials (BOM) and building information modelling (BIM). Other work, such as calculations, analyses and software, usually requires some supporting writing. However, much engineering documentation is in written form. Some common types of written engineering document are covered in Chapter 6.

Before we move onto the purpose of a specific document, we need to understand the high-level objective of the type of document. While all documents are broadly to communicate, this can be broken down into a handful of different reasons. Engineers often have to write reports, of which there are several types, for example, technical reports and progress reports, each for a different reason. Usually, each type of document will have the same rationale.

At this stage, we will explore some broad categories of rationale for engineering documents, which are to

- record
- inform
- instruct (or regulate)
- guide
- persuade (or influence)
- find out (or request).

Note that this is my interpretation of the main reasons why engineers write documents. Other rationales for communication may be added to this list, including a selection listed here. However, some of these, such as entertainment, are not for engineering purposes.

- Joan Nagle (1996) lists three categories of purpose, with a further breakdown of reasons for each
    - inform: instruct, notify, warn, interpret, clarify
    - request: apply for rights, obtain information, get action
    - persuade: make impression, influence decision, gain position, gain acceptance, recommend, sell.

- John Seely (2013) describes seven main purposes for writing and speaking: to interact, to inform, to find out, to influence, to regulate, to entertain and to record.
- Patrick Forsyth (2022) gives nine reasons for a report: inform, recommend, motivate, prompt or play a part in debate, persuade, impress, record, reinforce or build on existing situations or beliefs, or instruct.

Some engineering documents are produced simply to record something. They may be inputs to other documents, work or project deliverables. There may be a statutory or organisational requirement to record the information. Sometimes, these are filed away for reference and not used again (unless something goes wrong). At other times, they might be checked for regulatory compliance or quality assurance audits, or used for analysis, troubleshooting or input for future work. These are typically logging types of document, such as daily or change logs.

Informative documents communicate information; they inform the reader. They often document work done, such as research, a feasibility study, design, analysis and test. Examples include research papers and many technical reports. They usually have conclusions and may include recommendations.

An instruction controls or regulates a product or process, telling the reader precisely what is required or how to do something. These are broadly procedures and specifications, with unambiguous requirements. Examples include a standard operating procedure (SOP), standards, a statement of work (SOW), user manuals and maintenance manuals. However, there is a subtle distinction to be made between an instruction that must be followed and a guide. Hence, I suggest two separate rationales: instruct and guide.

Guides are similar to instructions but are not generally considered mandatory. Consider a quality management system (QMS), which comprises formal procedures. Alongside mandatory controls, there are often supporting guides, explaining how to do something or giving more of the background. These guides are not usually part of the QMS and are not subject to audit in the same way. Consider fault-finding guides: these may be helpful but do not have to be rigorously followed; there may be more than one way of doing something or the steps could be tried in a different order. Other examples are design guidelines, user guides and training manuals.

Persuasive writing is intended to influence the reader to do something, typically to spend money or agree a particular course of action. Although informative documents may make recommendations, these usually flow out of the report and, as such, persuasion is not the primary rationale. Typical examples of persuasive writing are bids and proposals. Persuasive documents need to make a convincing case.

The final rationale is to find something out. This may be exploratory, to see what is available in the marketplace, or to understand if something is satisfactory or not. Typically, this is the rationale with the tendering process, such as in a request for quotation (RFQ) or invitation to tender (ITT). With this rationale in mind, ITTs and similar documents often include a table to be completed, which means that the tenderer provides the required information in a consistent format. Another example is an audit questionnaire or indeed any type of form.

Some types of document may have more than one of these generic rationales. However, it is useful to understand the primary reason for any specific document. A technical report may be informative, for example, to document work done or a test completed, or to identify the root cause of a problem. Other reports may be persuasive, for example, to make a case to invest in new equipment or technology.

The rationale may inform the title or content of a document. For example, a user manual implies an instruction, while a user guide suggests that it is optional. An inspection & test procedure (ITP) is an instruction, with precise details of what to inspect and test. However, an ITP may have a secondary rationale, for example, to confirm that a product or building meets the stated requirements (or not). It may be useful, therefore, to include an inspection checklist where items can be ticked off, or a blank table to enter test results for demonstrating compliance.

When starting to write any engineering document, it is useful to be clear on the primary rationale and, if appropriate, the secondary rationale, one too. Six common rationales for engineering documents are to record, inform, instruct (or regulate), guide, persuade (or influence) and find out.

As an exercise, consider the following examples. What is the rationale of these documents?

1. Engineering design log
2. Progress report
3. Acceptance test specification
4. Request for proposal (RFP)
5. Proposal
6. Software release note
7. Audit report
8. Subject access request
9. Troubleshooting document
10. Business case
11. Onboarding documentation
12. Technical service bulletin.

See Table A.9 in the appendix for the solution.

2.2. Defining the purpose

Building on the rationale, defining the purpose of any engineering document should be part of the plan. It is one of the key aspects of effective writing, the P in PRIDE. Consider doing archery – if someone wants to hit a target, they need to aim at it. There is no point releasing the arrow if they have no sense of the direction, as it could go anywhere. Similarly, with writing any document, they need to plan what they are aiming to do before starting to write. The good news is that, like learning archery, with practice it gets easier to hit the target.

> The man who chases two rabbits catches neither.
>
> (attributed to Confucius)

The rationale will usually be the same for each type of document. For example, a test report will inform the reader. However, the purpose will be unique for every document, as every test will be different. Even if the same test procedure is followed, the test subject and results may be very different.

Ideally a document has only one purpose (and one target reader). A document that aims for several purposes may not hit any of them well. If there is more than one purpose, consider if more than one document might be more appropriate. For example, an engineer may need to give instructions on how to do something and also demonstrate compliance to a regulator. In that case, two documents might be better, with one referring to the other.

If a reader picks up a document and asks, 'So what? What am I supposed to do with this?' then the writer has failed to communicate their purpose. If the writer is not clear in their own mind about the message that they want to send, then the reader almost certainly will not receive it. This is only a small step in the writing process but can make a massive difference, as an engineer can then start to write with a clear purpose.

A simple framework of 'know, feel, do' (KFD) is widely used in communications. In preparing to write, ask what the reader or audience needs to know, feel and do, having read the document. This works for most communications, including talks and presentations, and is particularly useful for reports. These should be succinct bullet points. A supplemental question to ask is, 'What do they already know, feel and do?'

The 'know' is the message, ideally one but a maximum of three points to avoid diluting a key message. Engineers may think that the 'feel' is a bit soft and fuzzy, but this is a useful question. For example, feelings could be confident, satisfied, convinced or safe. This also illustrates the need for proof-reading: if a reader sees a document full of spelling, grammar and punctuation mistakes, are they then going to feel convinced by the technical calculations? The 'do' may be a specific action or simply agree, accept or continue.

The purpose should not be generic; it needs to be specific. 'Know what I did; do the recommendations,' could be written about every report. It is also useful to get a sense of scale or perspective. If a report is proposing that the reader spends a million pounds on new technology, they might need compelling arguments to be convinced. If there are audit findings, are they minor non-conformities or something more serious?

> *Know*. On routine inspection, a crack was found on the vessel that means it is unsafe to use.
>
> *Feel*. Convinced this problem needs urgent attention.
>
> *Do*. Repair the crack before the vessel is used again.

This KFD framework is a practical tool for the writer to understand the purpose before they start to write. It is not something that goes into the document itself. However, it is useful to write it down, as the very act of writing helps to clarify thoughts. It can be used in agreeing a brief (see Section 2.10), and as a reference when checking and reviewing the completed document.

Defining the purpose of any document is essential. Alongside that, a writer needs to identify the target reader. There is a link here, as defining the purpose involves considering what the writer wants the reader to do with the document. A writer should always view a document from the reader's perspective.

2.3. Identifying the readership and their needs

> You never really know a man until you stand in his shoes and walk around in them.
>
> (Harper Lee, *To Kill a Mockingbird*)

Alongside getting clarity on the purpose, understanding the reader is a key part of planning any engineering document, the R in PRIDE. This is linked to the purpose, understanding what the reader is going to do with the document, or at least what the writer would like them to do. A document is written for the reader, not the writer. Always consider the reader's perspective.

As a reader, what would we look for in a similar document? Having a good structure helps the reader find what they are looking for. This extends to putting the right content under each heading; for example, conclusions should be in the conclusions section, not the introduction (which I have seen). Pay attention to points of detail, such as a contents section with page numbers, which are like signposts that help the reader navigate a document. Good readability and presentation also helps them.

There are several general considerations about a reader, particularly around their level of technical expertise, level of involvement and role. For readers outside an organisation, there are also information control and commercial considerations. Some documents, such as company newsletters, may be primarily aimed internally at employees but are visible to external people when they are left in reception.

One of the first things to understand about the reader is whether or not they are technical, with a background in science or engineering. If they are technical, what is their level of competence, perhaps engineering technician or engineer? For example, writing a process specification (or any kind of written instruction of how to do something) may be very different if the reader has a specific competency rather than none. What is the reader's area of expertise? They might be a Chartered Engineer but work in a very different field, in which case they may not be familiar with the technical specialism.

Another consideration is whether the reader is familiar with the project. For example, when writing a report proposing something completely new, it probably needs the full background context in the introduction. However, if the report is on a work package that is part of the reader's project, then they already know all about it. A report should always cover the background in the introduction, but it would be much shorter if the target reader is already familiar with the project.

Linking back to the purpose, a writer needs to understand not only the reader, but also the context in which they are operating and therefore what information they need. For some documents, such as proposals and reports, a writer should know whether their target reader is the decision maker or not. If so, are they an engineer, who wants the technical facts, or the budget holder, who wants a costed business case for an investment?

If they are not the decision maker, but could champion the proposal, then they will need the facts and figures summarised in a way to help them do that. For example, if they are a council's planning committee, an application needs to demonstrate how the relevant planning criteria are met, and be presented in a way that the committee can defend their decision to the wider council or members of the public.

For proposals and reports, understand if the reader wants recommendations. Often they do but not always. Sometimes readers want to make an informed choice and would like to be presented with a number of options, with the potential advantages and disadvantages of each, so that they can select the best way forward themselves. They would not be satisfied if they wanted recommendations and these were not provided, or vice versa.

Finally, ideally, there should be just one target reader (or group of readers). The exception is within the reporting line; for example, a project manager might want to complete a project deliverable, and a project engineer reporting to them might use the detail in the document. If there is more than one reader, consider whether more than one document might be better, particularly if there is more than one purpose.

For publications, such as journals and magazines, the readership may be indicated by the title. Titles such as *New Civil Engineer* and *Professional Engineering*, the official magazines of ICE and IMechE, respectively, make it clear who their readers are. Journals such as *Advances in Cement Research* and *Infrastructure Asset Management* have more niche, specialist target readers. Most publications will have guidelines, which will include who their readership is and the reach or size of their circulation.

Consider how the reader is likely to receive the document, particularly if they might not like the message that it is giving them. For example, the research they have funded has proved inconclusive or the product cannot comply with a key technical requirement. In such cases, communicate with them face-to-face first rather than just sending them a written document.

Another key issue is whether the reader is internal or external to the writer's organisation, as this can impact several areas to do with classifying information, information control and the commercial relationship.

2.4. Classifying information

Engineers need to understand what information is required for a document to achieve its purpose and also what may be made available to their intended reader. This is often driven by whether they are internal or external to the writer's organisation. Larger organisations may even have information control between departments or divisions, due to their client base, such as in the defence sector. Sometimes they are providing consultancy services to clients who are potentially competitors. Other restrictions may be due to security, geographic location or personnel data.

Proprietary information may be technical, financial or business information. There are various types of intellectual property (IP). Some are protected by trademarks, registered designs and copyright and may need to be identified as such, for example using the copyright © or registered trademark ® symbols. An organisation may also have guidelines as to how a trademark or brand is to be used. Other types of IP, such as trade secrets, must not be disclosed to external parties. When working on a patentable novel invention, do not disclose it to a third party before filing.

Understanding at the planning stage what information may or may not be shared with the reader can save a lot of rework in the editing stage. If information cannot be shared, this might mean reconsidering the whole approach, such as 'make' not 'buy'. Note that readers may also need access to any reference documents. For example, a specification placed on a supplier may require compliance with other company standards. Some proprietary information may be made available to third parties under a non-disclosure agreement (NDA), which must be put in place before any information is shared.

There may also be commercially sensitive information from a competitive advantage perspective, such as internal costings or business plans. Suppliers may choose not to provide manufacturing drawings to a customer who only needs the assembly drawing for their purposes. In the same way, the supply of other written engineering documents may be limited to those agreed under a contract. Names of other clients and their confidential information must not be disclosed to third parties.

Another broader issue is the contractual situation. People may be more circumspect over the content if it is going to an external audience rather than just an internal one. This may be a matter of presenting things in a more diplomatic way, particularly if the client or business partner is spending money. There may be commercial implications if product or service capabilities are overstated. Writers need to be aware that even internal documents may become disclosable in a court of law at a later date, so anything they write may be discoverable.

There may also be specific legislation applicable for information control. The UK government has strategic export control for which an export licence is required. This covers goods, software and technology, which may include related documentation. The control list includes military, radioactive and dual-use, such as aerospace.

The General Data Protection Regulation (GDPR) has seven principles about controlling and processing personal data. The UK government's website (www.gov.uk) provides further details of UK strategic export controls and GDPR. The Information Commissioner can issue a monetary penalty for failing to comply.

As part of integrating writing with engineering work, information gathering should be managed. If content is to go into the document, is it a core part of the 'story' that needs to be in the main body, or will it go in an appendix as supporting information? Aside from the document retention, what other information is to be stored, in what format, where and for how long?

Source references for input facts and data are needed as evidence to substantiate findings and conclusions. Complete citation details need to be captured at the point of use, to avoid repeating effort. For online sources, this includes the URL and the date accessed, as content may change over time. Traceability of any information is particularly important in the case of specifications and the commercial interface.

2.5. Commercial and marketing interfaces

In most projects, there are multiple documents. For a large project or contract, there may be very many. While engineers are often focusing on technical documentation, they may be contributing to or interfacing with other documents. It is important to understand the interfaces between the technical, commercial and marketing elements in the suite of documents and how they fit together in the business context. Information should be consistent across different documents and there should be a clear flow between them. Document change should also be controlled.

One way to picture this, as an engineer, is that documents may be in parallel or in series. If they are in parallel, then they coexist at the same point in time in the project. For example, instead of one very large product specification, there could be one overarching product specification and several individual component ones. This makes sense, as they are for different readers (who may be internal or external) and purposes (focusing on different parts).

If documents are in series, then they will flow sequentially, with one document becoming an input to the next. For example, there may be an increasing level of detail in a series of specifications, or a new document for each phase in a gated project process. Consider a customer giving an order to a company who sub-contracts part of it. The company still has the responsibility of ensuring that the whole product or service comes together and works as intended, and will need to demonstrate compliance to their customer. The company, therefore, needs to flow down the relevant customer requirements, by including them in the documentation that they in turn pass on to their sub-contractor. This includes requirements for the product itself and the associated verification documentation, such as drawings and material certification.

A contract is a legally binding agreement requiring an offer and acceptance. In law, a contract may be verbal or written. It is good practice in engineering to have a written contract and only allow changes in writing by an agreed process with a nominated person. If the contract is silent on the issue, then it may not prevent a verbal change to contract or an unauthorised person agreeing a change. A few words in the contract (or specification) to say that any changes must be agreed in writing with the project manager (or other nominated role) could avoid problems later on.

Consider a scenario where a company placed a contract with a supplier to provide a service on site. Once they started, they found a problem and proposed an alternative solution to someone on site, wearing a hi-vis jacket with the company name on. That could have been anyone, including the cleaner. A casual comment, such as, 'That looks OK to me,' might be deemed acceptance of a change. Everyone on site should have a clear understanding of their roles and responsibilities. I heard one example where an engineer was genuinely trying to be helpful and agreed to site changes. However, it was not within their remit, and they were not aware of the full project context.

It may be that such a proposed alternative solution is not acceptable and could require rework, with associated delays and extra cost. If any representative of the company had agreed the change (and the contract was silent on the issue of who could authorise that), then it may well be at the company's cost to fix it, not the supplier's.

Even with the correct authority, changes agreed only in conversation may result in several problems: a disagreement on what was actually agreed; the 'paperwork' being delayed in being updated or maybe never getting updated at all (so 'as-built drawings' are not correct); or a late unexpected extra invoice because there was no record of the change. Hence, the importance of ensuring any changes are to be agreed in writing with a nominated point of contact.

Another important issue is to define the order of precedence in contract documentation. While consistency is important in writing, the more documents there are, the greater the risk of inconsistency. Good practice is for the contract (or a purchase order) to state the order of precedence, and therefore it is clear what should be followed in the event of any conflict.

Marketing collateral may be less technical or specific, compared with many engineering documents. While these may be perceived as having a more positive spin on products and services, they should still be consistent and not misleading, particularly in implying capabilities. Though not forming part of a contract, they may raise customers' expectations and could damage a company's reputation.

2.6. Technological tools to support writing

It is only on rare occasions now that engineers need to sit down and write an essay or other document by hand. The most common technological tools to support engineering writing are word processors or word editors.

Microsoft Word, part of the Microsoft Office/Microsoft 365 suite, is the dominant player in the market. There are also Google Docs, WPS Office, LibreOffice and many more software options, including free software, although often free versions have more basic features. There are more specialist editing tools, such as Grammarly, ProWritingAid and Overleaf (a LaTeX editor), and further writing tools for publishing.

All software has different features, pricing structures, advantages and disadvantages, and some are better than others in different areas. It is not within the scope of this book to evaluate all the available options, nor is it within my expertise to do so. As an engineer and writer, I have predominantly used Microsoft Word in my career. I will highlight some of the features of Word that I (or I have seen that other engineers) have found particularly useful. Other software may have similar functionality. The version of Microsoft Word that I am using is Microsoft 365 apps for business (updated in March 2026).

I have often found in delivering training courses that engineers are not aware of some of the useful functionality in Microsoft Word, in particular the readability statistics (see Section 5.1). Also, in using Microsoft Word for proof-reading and editing, that it can check not only for spelling, but grammar and punctuation too. It can also check for biases, such as gender or ethnic bias, and many aspects of clear and concise writing style. However, these options need to be selected in the settings.

From the Review tab, select the Language drop-down box, then Language preferences. From that Word Options dialogue box, select Proofing near the top left. From that dialogue box, lower centre, select Grammar & Refinements from the Writing style drop-down box, and click on the Settings button to the right to see (and select) the full list of that Microsoft Word can check for. These points of detail include grammar, clarity, conciseness, formality and inclusiveness. Read Aloud is another useful function on the Review tab for proof-reading.

The dictate tool, found on the Home tab, can help save time, as engineers can rarely type as fast as they talk. This is particularly useful when doing an initial brain dump of ideas. Microsoft Word will capture what is said and then it can be tidied up afterwards. It can also help when trying to explain something because engineers are more likely to use simpler, everyday words that they would speak. Imagine sitting next to someone and talking to them. However, it does need to be checked, as although the dictate function is very good, it is not perfect. I once said hierarchical (in the context of a project work breakdown structure), which was captured as 'hey how cool', while 'scrum master' was captured as 'school master'.

Microsoft Word has the citations function under References. From the Insert Citation dialogue box, it prompts the writer for the fields required for each reference type and then presents them in the correct format. References can also be copied from the user's master list to the current list, so references added for previous documents do not need to be re-entered. This is an example of a Word tool that ensures that something is correct and complete and saves the writer time. However, citations need to be entered manually the first time.

Citation or reference management software can be useful, particularly in research, where there are many citations from journals and other academic sources. These can directly import the citation into a document, rather than having to manually enter each field for author, title, publication and so on. Examples of reference management software are EndNote, Mendeley and Zotero. While academia and research institutions may have many references, in industry there tend to be fewer, often internal documents, which are not in a retrievable format as citations.

Aside from the word processing, editing and reference management, other software may support writing. Document management systems are used for storage and retrieval, with version control to manage revisions. They are usually configured with keywords and other metadata, to help search and retrieve documents.

Document management systems are sometimes used during the approval process (see Section 3.9). However, as this typically involves tracking document changes for all corrections, it can be time-consuming and may make a document look very messy. I would recommend that documents are only submitted into the system when the formal review process starts, after proof-reading and the associated editing has been completed.

Engineers may use other software in their work, some of which may be related to writing. For example, in requirements management (see Section 6.2), software for writing and checking requirements or a requirements management system (or database).

As for any tool, choose the best software tool(s) for the task in hand. These are usually determined by the engineer's company or organisation, who may restrict the use of other software. Some individuals prefer one editing tool but then need to switch to their organisation's adopted tool to issue something. This can be inefficient as it is extra work and there is a risk of formatting not carrying over.

Within an organisation, technology tools should be integrated where appropriate and consistently used by everyone, and there should not be too many of them. More software tools increase subscription costs, potential incompatibility issues and the time for users to become familiar with them. There are also inefficiencies where multiple systems have to be monitored and sometimes duplicate effort.

Engineers often communicate in writing through many channels, including emails, Microsoft Teams and messaging. Important information can get lost in an email or other communication. Teams working together should have clear communication protocols and a common understanding of what tool to use for what purpose. For example, the 'office WhatsApp' should be used for social purposes only. There should be guidelines for the use of communication systems, such as workflows, technical queries and task management systems. While I am focusing in this book on writing engineering documents, the same guidance on writing well applies to all written communications.

2.7. Artificial intelligence

Artificial intelligence (AI) can also support writing. There is a risk in writing about AI, given the current pace of change, that this status will be obsolete before this book is even published. To a certain extent, AI is embedded in many of the editing tools covered in Section 2.6, such as the dictate function.

Voice recognition systems have come on greatly in recent years. I recall buying Dragon NaturallySpeaking speech recognition software around 2011, when it was state of the art. I had to 'train my dragon' by reading a set list of words and even then it was only trained for my voice. It would confuse words like 'project' and 'budget'. Even so, it was a very useful tool for capturing input and saving me time. Microsoft acquired Nuance Communications Inc., who had the rights to the Dragon product line, in 2022, and now Microsoft Word has a built-in dictate function.

Other tools, like Microsoft Copilot and ChatGPT, are more AI tools with great potential, for example in drafting, rewriting or summarising documents.

> The current state of generative AI technology represented by ChatGPT is impressive but flawed.
>
> (Qadir, 2023)

AI tools can produce quality written documents instantly, hence saving a lot of time. They can also write in grammatically correct English, helping those for whom English is not their first language with proof-reading corrections. I know many engineers who have embraced the use of AI and claim huge benefits in time-saving and quality of writing. Despite these benefits, there are risks and limitations with AI.

To understand these risks, we need to understand what AI is in the context of writing. ChatGPT is a natural language processing (NLP) model based on Generative Pre-trained Transformer (GPT) (Qadir, 2023). Microsoft Copilot is a similar GPT-based chatbot and is integrated with Microsoft 365 apps. They both use a large language model (LLM), which is 'trained' on a large volume of data. There is a risk of bias where a model has been trained on a limited dataset.

There is a lack of traceability to source data, and therefore a risk of breaching others' intellectual property rights (IPR). There is also a security risk to an organisation's own IPR and other sensitive data. Unless a protected private network is set up, the tool will learn from any input, thereby effectively potentially putting an organisation's information into the public domain.

AI output cannot be considered a writer's original work, which may mean plagiarism, and so its use may not be permitted in universities for writing assessed work. Where AI is permitted and has been used, such as writing an article for a magazine, it is appropriate to declare that (see Section 3.6).

These models also have interfaces with the human language and computers, leading to the term 'prompt engineering' to describe the human interface. This is another skill to be learnt: how to ask for what is needed with appropriate commands. The right questions need to be asked to get the required answer.

While AI-generated documents may read well, as they are written in good English, the content may be nonsense. Another risk of AI is what is described as 'hallucinating'. In 2025, Deloitte repaid AU$440000 (£230 000) after admitting (in a new version of its report) that it had used generative AI in its report, which made up a quote and several non-existent citations (Page, 2025).

> Even with ChatGPT writing for you, being able to write yourself is the most important skill of the next decade. AI gives you answers. Writing gives you understanding. Understanding creates better questions. The person asking better questions owns the future.
>
> (Bartlett, 2026)

AI generally needs to be used responsibly and for appropriate purposes. AI is not software and cannot be treated as such. While AI can be a great support tool, it cannot be relied on without careful checking. It can make mistakes and even hallucinate. Writing still needs to retain the human element. To be able to technically check and correct AI output, an engineer needs to fully understand the subject and context themselves. They need to know what good looks like and when the written output is wrong.

Technology in itself is neither good nor bad. There are many benefits to AI technology, to be balanced with the ethical and accuracy issues. The key is to understand the benefits and limitations and use the tool appropriately.

2.8. Structure and templates

Structure builds on understanding the type of document, purpose, reader and business context. This is a key step in planning any document. Even if there are no headings, such as with an essay style, there is still a logical structure or order for the content. Structure helps writing to flow, like a story, with a start, middle and end. Alternatively, it might be a logical organisation of the material, whatever best fits the purpose. Once the 'skeleton' of the structure is defined, it is easier to 'flesh out' the required content in the drafting stage. Structure gives the framework, following the purpose and scope, so that the required content, and that only, is included.

A structure is not the same as a template. A structure is important and covers the headings; a template includes the structure and much more. A template is an electronic file from which new documents are created. There are many reasons for developing and using a template. As we will see in Chapter 4, these are mostly aspects of being clear, correct, complete and consistent. A template gives a consistent, professional presentation and addresses accessibility issues (see Section 5.9). Here are ten reasons for using a template

- correct headings for document type
- no important content missed
- avoid irrelevant content
- consistency
- save time
- avoid errors from cut and paste
- settings (proofing language, font and format)
- footer disclaimers and copyright information
- correct company information
- accessibility.

Sometimes, organisations have two levels of template. A master template covers the general company information, font and style details for consistency. From that, individual templates are created for each type of document, with the relevant section headings. In that way, all documents across the team have the same look and feel. Be aware that external clients may require that their template is used. If not, they may still have requirements, such as proofing language that needs to be incorporated.

I recommend always using a template, particularly as engineers often use only one or two document types for the majority of their work. Where there is no template, create one with the section headings and save it in template format before starting to add any content. Even if there are no headings, still follow the process of having or creating a master level template, as the template does so much more than headings. Then outline the structure, which may just be prompt words in normal text.

In Microsoft Word, the template remains unchanged if a new Word document is created from it. It is saved as a template format file (*.dotx), not a document (*.docx) file, and in the user's templates folder. To find this templates folder, select File, Save as, select Word template format, and then a folder should be recommended.

Use an automated table of contents, including page numbers, with a page break before and after. Choose a numbered heading style as this helps the reader. These points of detail allow the reader to navigate the document and find what they need. Check that the heading style is correct for each level. Note that using headings with Microsoft styles is more than a different font or size; it is part of the document outlining and helps with accessibility, as screen readers can distinguish headings from body text.

An accessibility check of a template can reduce the number of items flagged in checking accessibility for all documents created from it. I recall one template with a logo, with no 'alt text' (see Section 5.9), not even simply 'logo' or labelled as decorative text. Every single document created from it flagged the same logo every time in the accessibility check.

Try to avoid too many levels, ideally a maximum of three, as it gets hard to read with something like Section 5.3.1.7. If it is a large report, consider if more than one might be appropriate. Very large documents may present extra challenges, not least being daunting for the reader. From a document management perspective, it may also make sense to break a large file down into a number of smaller ones.

Smaller documents are quicker to review and may need fewer checkers and approvers. Future version control is also easier should revisions be required. When a document is revised, it is raised in issue, for example from issue 1 to issue 2. It is implied that the whole document has been checked and is still current and correct at the point of reissue. This may mean a lot of work for a small change, or it goes into the 'too difficult' category and the document never gets updated.

Where there is an existing template with numbered sections, and one section is not required for a specific document, it is preferable to maintain the standard structure to help the reader, as they may be familiar with and look for particular sections. For clarity, rather than leave it blank, add a comment such as 'Section 3 intentionally blank', so that the reader does not think that something is missing.

In the absence of a template (and even when there is one), engineers sometimes go back to a previous example and start with that. While it is useful to have good examples available, this is not best practice. The structure may be different; for example, the last report deleted a section of the template that was not required, or it was written on an old version of the template. Cut and paste often leaves telltale signs, like mixed proofing languages and odd changes in format. There is a risk with following a previous example, in that it is not a good one, at least not for the particular document required. Make it a habit to always go back to a clean functioning template.

In the absence of a style guide, a template may also provide some style guidelines and notes. These may include examples of how to present information, such as content to put in a summary or referencing style. *Lorem ipsum* filler text can be used in a template to show text placement and format. Guidelines or examples of standard text, which the writer later deletes, may be provided in italics or as hidden text in the template.

In Microsoft Word, 'hidden' text can be set to display on the screen but not print. Hidden text is accessed from the same menu that displays superscripts: highlight the text to be hidden; then, from the Home tab, launch the Fonts dialogue box (with the arrow in the bottom right of the Font section); then check the box under Effects saying Hidden.

To display hidden text, select the File tab, then Options (at the bottom left of the screen), then Display and check the box that says hidden text. On the same Display dialogue box, deselect the printing option to print hidden text. Hidden text is then visible on the screen, shown with a light dotted grey underline, but not visible when printed (or printed to pdf format). Tip: be careful not to use the captions feature if showing how to present figures in hidden text in Microsoft Word, as automatic numbering continues even if the figure is hidden.

Table 2.1 gives a checklist for the main points to consider in creating a template. Table A.1 gives an expanded template checklist, covering details such as captions and bullet points. Prepare the template and draft the document in the same software editing tool, such as Microsoft Word or Google Docs. If not, it can take a long time to convert it later, as formatting may not copy across. An appropriate structure on a clean, functioning template is a key output from a plan.

Table 2.1 Template checklist for Microsoft Word

Item	**Checked**
Title and organisation information	
Numbered headings and sub-headings in themes/styles	
Automatic table of contents	
Lists of figures and tables (if applicable)	
References section	
Header (or footer) with document number and revision	
Glossary of acronyms & abbreviations	
Footer with page number, copyright and confidentiality notice	
Proofing language: UK English	
Sample text identified (if used): italics/hidden/*lorem ipsum*	
Paragraph settings: widow and orphan control, do not hyphenate	
Accessibility check	

2.9. Identifying scope and what to include

With the purpose defined, the reader identified and the structure (and template) completed, there should be a good framework to prepare to write. The structure itself covers what the document needs for it to be complete, and therefore what to include and what to omit. If there is no heading, no content is required; if there is a heading, content is required. For a simple document, this may be sufficient.

Writing should be fully integrated with an engineer's work, not an afterthought or bolt-on at the end. This is the I for Integration in PRIDE. Engineering writing is rarely a stand-alone activity, but part of other work, such as research or design. Planning for a project should include the writing. Getting clarity on the brief for the document may mean getting clarity on the scope of the work.

Ideally, the person requesting or commissioning a document gives a brief to the writer. Unfortunately, this brief (as in a summary) is often brief (as in short)! In that case, the writer may need to clarify the brief in developing the document plan (see Section 2.10). In particular, the plan needs to capture any particular requirements about the content, such as 'must include a site plan'. It is also helpful to indicate the depth required with a guide to the size of the document, such as '5000–10 000 words' or '10–20 pages'.

The scope of a document sets the boundaries of what is to be included and, equally important, what is not. For example, is a procedure for use on a particular site or all the company's sites? Any misunderstanding around scope could lead to an incomplete document or significantly more work. Defining scope well helps avoid 'scope creep'. The contractual scope of supply is needed to write a specification. For example, does it cover the product delivery only or include installation and commissioning? Any limitations should also be noted, such as no budget available beyond the writer's time or limited access due to security.

Any related documents should be identified upfront and fully referenced. This may be previous work or commercial input. This applies to doing the work as well as the document itself. For example, a design report may require compliance with a particular specification and standards. If that is not done from the start, then the design may be rejected.

Sometimes engineers do the work first, then write it up at the end, typically for smaller documents. At other times, they may be writing up section by section as they progress their work. This tends to be for larger documents, where the document is the main deliverable from the work. It can also be useful for knowledge capture in larger projects.

The scope of the work, and therefore the document, may not be fully defined at the start of the project. For example, an academic assignment typically starts with a broad topic. A literature review then covers the available research and identifies the gap in the knowledge for the research to focus on. In industry, an engineer might have a brief to research how AI could be used in their company. After some initial work, they will focus on a few areas.

To a large extent, organisation of ideas is part of the writing stage (see Section 3.5). However, a certain amount has to be done in the planning stage to determine the document structure.

For example, is the work to be presented chronologically or in another way? The scope and content requirements need to be sufficient to complete the planning stage. This is a key part of a document plan and provides a framework to proceed to the writing stage, without the risk of wasting time and effort.

2.10. Developing and refining a master plan

Best practice for writing any engineering document is to construct a plan before starting to draft the content. A good document plan covers the rationale, purpose, reader, structure and scope of the document, as already covered in this chapter. In developing a complete plan, the writer should also engage with the relevant stakeholders and estimate the timescales. An engineer should know if there are any procedures or guidance notes about issuing documents. Also, they should ensure that they (and anyone else who needs it) has access to the relevant document management systems.

In many organisations, the person requesting a document is the same person who will be approving it, typically a line manager. However, in other cases, this is a different person. For example, a project manager requests a report and a technical authority approves it. Either way, once the plan is complete, the writer should seek out the approver, have a conversation and agree the plan with them before proceeding to the writing stage. This is particularly important when the person providing the brief is not the approver. Otherwise, the approver may not be aware of the document's existence or have any opportunity to be involved in its creation.

Writing down a plan helps the writer to clarify their own ideas and better communicate with their approver to get agreement. If necessary, refine the plan based on this discussion. Agreeing the plan with the approver confirms the writer's understanding of the brief (or expands on it if it was incomplete). It is also an opportunity to discuss the amount of effort involved, and both parties can commit to realistic timescales, avoiding the stress of mismatched expectations. Getting buy-in for the plan can save time and effort for both the writer and approver because expectations are aligned upfront, so any comments and editing in the review stage are minimal.

I have heard people say that they do not know who will approve their document until they have finished writing it, or that the approver could be any of a number of people, whoever happens to be available at that point. Neither of these situations is ideal, as the approver should be consulted at the planning stage. If input is needed from others or there are multiple authors, understand how that is going to work upfront (see Section 3.4). Ideally, talk to everyone in the document creation and review process at the planning stage. It is always good to engage with stakeholders early, and they may have specific requirements to be aware of. This can be particularly useful if an engineer has not worked with them before.

It is important to understand who will be involved in the review process at the planning stage and to clarify the roles and responsibilities of each person. The term 'reviewer' is often used quite broadly and can cover a range of roles. Sometimes writers ask a colleague vaguely to 'have a look' at their document. Define the reviewing roles, what is meant by them, and have a checklist of what each role is to review.

Incidentally, the word 'proof-read' or 'proofread' may be spelt with or without a hyphen. Either is acceptable. Personally, I err on the side of using extra punctuation if it helps to make it clearer. As for any point of style, consistency is key, so where there are possible variations like this, adopt one form and always use that.

Reviewing and approving is covered in more detail in Section 3.9. At the planning stage, the important thing is to identify roles and responsibilities. For formal documents, writers do not approve their own work. Ideally, there will be at least three people involved and three sequential stages of checking, reviewing and approving a document.

Some organisations refer to proof-readers, others to editorial checkers for the same scope; some have technical checkers and reviewers, others have reviewers and approvers. It does not really matter what they are called but, as for many other aspects of writing, an organisation should be consistent and have a common understanding of what it means by that role and who is checking which points of detail. There should only be one role with primary responsibility to look at each point of detail, not potentially three or more people all checking for the same thing or potentially all omitting a check. I will use the terms proof-reader, technical reviewer and approver here.

There may be several technical checkers and occasionally several approvers. Depending on the organisation and context, the technical reviewer or approver may be a budget holder or other stakeholder, or someone with specific authority, specialist expertise or competence, such as health and safety. In the UK nuclear power industry and in defence, the term 'suitably qualified and experienced person' (SQEP) may be used.

Some organisations have formal written procedures for reviewing and approving documents, with defined roles and responsibilities and supporting checklists. At the extreme end of the scale, some organisations require marked-up documents or have formal written comment sheets that need to be completed for all comments, including proof-reading. I would not recommend having proof-reading as part of the formal review and approval process (see Section 3.9).

This section is focused on reviewing documents within engineering organisations. Outside of this, there are some differences in the review process and the roles. For academic qualifications, assignments should be entirely the student's own work. For journal articles, there will be peer reviews. For published documents, such as books, there may be additional roles, such as editor, publisher, typesetter and professional indexer. See also reviewing and approving (Section 3.9) and writing for publication (see Section 6.5).

No plan would be complete without an estimate of the time required. Writing should be planned alongside the project or other work. Sometimes people plan out their work and omit to allow time to write it up, or it is a bolt-on extra at the end. Writing is an integral part of the engineering work – note the I for Integration in PRIDE. It is how work is presented, so enough time should be allowed to do a professional job.

If input is needed from others, plan it early and get agreement from them as to when they will provide their information. Allow sufficient time for the review and approval process.

Reviewers and approvers are typically more senior people and may be busy. It is not fair or reasonable to present someone with an urgent document to review or approve just before a project, contractual or other deadline and expect them to drop everything. Again, this can be part of the buy-in conversation with the approver, understanding what a reasonable response time would be once the document is submitted. For example, they might estimate it would take them an hour to review, and they agree to respond within two working days.

The key point is not to start drafting without a plan. It does take a little time, but it is not much compared with the overall writing timescale. Moving the effort upfront into the planning stage brings clarity to everyone involved and means less work later on. The objective is to have a single pass through the plan, draft, edit and review process. This should produce a quality document in the least amount of time.

Having a time plan and monitoring progress against it can also help with improving an individual's writing skills and process. They can see what is taking most time and work

Table 2.2 Plan checklist

Item	Comment/completed
Rationale	Record/inform/instruct/guide/persuade/find out
Purpose	Know: Do: Feel:
Reader	
Circulation	Internal only/external use
Template to use (and location)	
Structure (if no template or need to adapt)	
Content requirements (if applicable)	
Size (approximate number of words/pages)	
Other related documents (if applicable)	
Style guide (if applicable)	
Software to use (and AI permitted or not)	
Time plan	
Author (lead author)	
Other authors (if applicable)	
Proof-reader	
Technical reviewer(s)	
Approver(s)	
Other notes (document management system/ approval process, boundaries, limitations, information controls, referencing style)	

on that. For example, if the writing takes a long time, work on drafting then editing (see Section 3.2). Reviewing actual timescales against the plan means that better estimates can be made for future documents. With a plan, a writer could give the reviewers and approver a heads-up on when they might expect to receive the document. Timescales for written documents can vary enormously. Typically days or weeks, it can be only hours or even months.

The '5Ws and 1H' is a useful tool for describing a situation or defining a problem. It is also a useful quick checklist for a document plan. Key parts of a plan are: why (purpose), who (reader), when (time plan). A plan should also cover: what (content, template), where (document management system) and how (review process).

A more comprehensive summary of what a document plan might contain is given in Table 2.2. This can be used as a checklist and for discussion with the approver before starting to write. The most important points in planning are the purpose, reader and integration of writing and work. The next chapter, on writing, will cover the drafting and editing of the acronym PRIDE for effective writing.

REFERENCES

Bartlett S (2026) https://www.linkedin.com/posts/stevenbartlett-123_unpopular-opinion-writing-is-going-to-be-activity-7422322603048128512-Mn3v/ (accessed 29/01/2026).

Billingsley J (2017) *So What? Effective Writing for Engineers*, 1st edn. CreateSpace, Scotts Valley, CA, USA.

Forsyth P (2022) *How to Write Reports & Proposals: Create Attention-Grabbing Documents That Achieve Your Goals*, 6th edn. Kogan Page, London, UK.

Nagle JG (1996) *Handbook for Preparing Engineering Documents: From Concept to Completion*. IEEE Press, Piscataway, NJ, USA.

Page C (2025) Deloitte refunds Aussie gov after AI fabrications slip into $440k welfare report. *The Register*, 6 Oct. https://www.theregister.com/2025/10/06/deloitte_ai_report_australia/ (accessed 05/02/2026).

Qadir J (2023) Engineering education in the era of ChatGPT: promise and pitfalls of generative AI for education. *2023 IEEE Global Engineering Education Conference (EDUCON)*. IEEE Press, Piscataway, NJ, USA.

Seely J (2013) *Oxford Guide to Effective Writing & Speaking: How to Communicate Clearly*, 3rd edn. Oxford University Press, Oxford, UK.

Technical Engineering Writing: A Practical Guide for Professional Engineers

Jean Billingsley
ISBN 978-1-83662-927-6
https://doi.org/10.1108/978-1-83662-924-520261004

Chapter 3
The writing process

The writing process starts with planning (see Chapter 2). Starting to write without an agreed plan is not good practice, and may result in considerable rework or even worse an unusable document. Before starting to write, draw up a plan, including clarity on the purpose and the reader. Another key deliverable of a plan is the structure. Think of this as the skeleton of a document, and writing is then about putting flesh on the bones.

Working with a structure helps keep the writer on purpose, as the required content is identified and unnecessary content can be avoided. Editing content out later wastes time twice, in writing it and then deleting it. If unnecessary content is left in, it adds no value and makes the document longer than it needs to be. Writing should be complete – enough but no more.

Start with bullet points under each heading. Consider what items need to be covered in a particular section of a document. Jot down just words or bullet points first, then put them in the best order (see Section 3.5). These points can then be expanded into paragraphs during the drafting stage.

At this bullet point stage, decide what figures and tables need to be included. These can communicate a lot of technical information and often in a better way. As the saying goes, 'A picture paints a thousand words,' so fewer words are needed if there are visuals (see Section 5.8). It may be appropriate to have the buy-in conversation with the approver at this stage, with the bullet points and figures outlined, rather than on completion of the plan. This may depend on the complexity of the document and the writer's level of experience.

3.1. Efficient and effective writing

An engineer's writing needs to be both efficient and effective. Efficient is defined as 'functioning or producing effectively and with the least waste of effort; competent' (Collins, 2023). Effective is 'productive of or capable of producing a result' (Collins, 2023). Productive use of time starts with a plan, then following a good writing process (see Section 3.2). Efficient and effective writing therefore means least effort and a quality result.

As with any engineering process, the aim should be for a single pass through the process and for the output to be 'right first time'. Going back to rework any step is a waste of time and effort and is not efficient. A good approach is also effective in producing the right result, a document that does what it needs to do. For example, getting clarity on the purpose helps the writer avoid going off on a tangent and writing about something that is not connected to the objective. A good writing process helps in writing quality documents in a short amount of time and is, therefore, both efficient and effective.

An environment that is conducive to writing also improves efficiency. That includes a physical space to work and a clear, uncluttered desk. Avoid all distractions not just from other people: close email and silence software, phone or other notifications. Put phones and devices out of reach or in a drawer to remove the temptation to keep checking them.

Avoiding distractions is a good time management tip generally, as allowing interruptions is a very inefficient way of working. With writing in particular, engineers need to concentrate on what they need to say. Even if they do nothing with the email they just scan read, allowing the distraction can mean losing their train of thought. Changing between different kinds of tasks can also be draining, as switching mindsets takes both time and energy.

> The most effective way to do it, is to do it.
>
> (Amelia Earhart)

From a practical perspective, find a space and time with the least interruptions to focus and block it out. Protect an hour or two for doing the most important work – writing. Individuals should recognise their own energy levels and when they have most energy. This is often the first 90 minutes of the day, although it is not the same for everyone.

Background music or sound can help with concentration. Again, individuals have their own preferences. When I am writing, I like to have background instrumental music (with no words to distract me), for example, movie soundtracks, classical music (such as Rachmaninov's piano concertos), Solfeggio frequencies or other focus music.

Finally, it is important to have a comfortable seating arrangement, good lighting and a well-designed workstation. The UK government's Health and Safety Executive (HSE, 2025) issues guidance for working safely with display screen equipment (DSE). The Regulations are applicable to 'workers who use DSE daily, for continuous periods of an hour or more', which probably covers most engineers.

3.2. Drafting and editing

As we have seen, starting to write without a plan is neither efficient nor effective. Chapter 2 introduced the PRIDE acronym for writing and covered purpose, reader and integration. This section builds on this, with draft and edit. Even with a plan, do not try to write something perfectly the first time.

Imagine that an engineer needs to cover three points in a section. They start writing the first one but get distracted by a red squiggly line that tells them that a word is misspelt, so they go back and correct it. Then they think that what they have just written does not read very clearly, so they reword it. Twenty minutes later, they are still stuck on that same paragraph, staring at the screen and have completely forgotten what their second and third points were. I often find engineers can relate to this scenario, which reflects a very inefficient and ineffective process.

The best way to write is to follow two clear stages: draft then edit. Drafting is about getting the rough content down, then editing is improving it. A good way to think about this is that the writer drafts for themselves and then edits for their reader. In this way, drafting and editing mirror the purpose and reader of the PRIDE approach.

The drafting stage addresses what the writer needs to say, the ideas they need to communicate, and is just getting down a brain dump of the key points and information needed. Ignore spelling and grammar mistakes (turn these off in settings if they are a distraction – just remember to turn them back on again when editing). Then the editing stage improves the presentation of that content to help the reader better understand it. It includes revising wording to be clearer, and also includes checking for accuracy and an appropriate style (see Chapter 4).

A large document might be drafted then edited by sections. For example, draft one section one day, then the following day, edit that section before moving on to draft the next one. The key point is not to draft and edit at the same time. Following a process of plan, draft and edit is an efficient and effective way to write. It also means that writing flows better, as there is a high-level document structure, so the sections flow, and starting by mapping out bullet points means that the paragraphs flow too. Draft then edit is one of the top tips I would give engineers when writing anything.

3.3. Dealing with writer's block

Some writers say that they get writer's block; others say that there is no such thing, and it is just procrastination. Certain types of writing, like novels, require a creative spark, developing plots and characters. However, engineering writing should not be creative in that sense, but setting down factual and objective content, so engineers should never have writer's block.

If an engineer feels that they are 'stuck', my suggestion would be to take a step back and understand why. Break it down: where are they in the writing process (plan, draft or edit) and what is the specific problem? If it is a question of not knowing what is required or where to start, review the brief and the plan (see Section 2.10). If the document plan is not complete, go back and finish that. Maybe the work itself needs to be completed first.

If there is an agreed plan, understand whether the block is a question of not knowing what to say or not knowing how to say it, in other words, is it drafting or editing (see Section 3.2). If it is the former, start with the first section in the structure and jot down single words or bullet points (see Section 3.5). If it is a question of having a bullet point, but not knowing where to begin, try using the dictate function in the word processor to capture it. Speak as though explaining the point to someone sitting at the next desk. If something written does not read well, analyse whether it makes the point or not. If not, try and explain it again.

If the point is captured but not phrased well, then it is a question of editing. Perhaps a sentence is too long or unclear. Try taking a break and coming back to the draft. Chapter 4 covers writing style and some common issues that engineers face, such as finding the right word, wordy habits and avoiding ambiguity.

Engineers also need to create the right environment and avoid distractions (see Section 3.1), so that they can immerse themselves in writing. If there are other issues, find a time or place to avoid interruptions. When starting to write, allow a few minutes to sit with it and settle the mind. Sometimes, it can take a while to get into 'flow'.

When I am really immersed in writing, I like to think of it as being 'in the zone'. However, it may not be easy to get into that state of mind. Stick at it for 20 minutes or find a simple task to ease the way into writing. Perhaps one task to complete today might be to find the right template and create a new document ready to start writing tomorrow.

Sometimes, if an engineer has something else on their mind, they need to stop and deal with that: either make some progress on doing it or make a note to do it later. If it is a matter of having no energy, take a break or come back to it fresh the next day. This is another good reason for planning and integrating writing into a work plan. There should be a margin of time to allow some flexibility and avoid the pressure of writing to a deadline.

3.4. Writing with collaborators

Engineers often work collaboratively on projects with partners, ranging from a university research project to designing planes. In project management, a useful tool is a roles and responsibilities matrix, to be agreed on at the start of a project. There are different versions of these matrices, but a key issue to clarify is who has primary responsibility for each aspect of the project. This also extends to writing and making a distinction between others providing input to a document and being a co-author.

Often, engineers are the sole author of a document, with other people only involved in the review and approval process (see Section 2.10 and Section 3.9). At other times, there are multiple authors involved in one document, such as an academic paper. It is important to agree upfront who is responsible for what and how everyone is going to work together, which may be described as a 'who's doing what' checklist (Lindsell-Roberts, 2023). Ideally, this should cover not only roles and responsibilities but also agreed writing style, software tools and protocols for working together.

Problems can arise with inconsistencies when there are multiple authors, for example within a suite of contract documentation (see Section 2.5). Even when an order of precedence is specified, it can lead to unintended consequences, such as when a weaker requirement takes precedence over a more prescriptive one.

Documents written by multiple authors may also have a noticeable change in focus or writing style. Even if it is not apparent to the reader that this is due to multiple authors, the writing may not flow well, due to inconsistent writing style, so the reader may find it harder to read.

It is more likely that there will be 'errors and omissions', a term commonly used for mistakes and oversights, when there are multiple authors. There is a risk that no single author feels ownership of the whole document, and some may make assumptions about what the other authors are covering.

Where people are working collaboratively on the same draft document, software tools and protocols for working together need to be agreed, such as using read-only or editable files. For example, Microsoft SharePoint can be used for working together and sharing documents. Are people collaborating in real time on the same version or can only one person modify it? People may lose their sense of ownership (and errors and inconsistencies may be introduced) if someone else edits their work in progress.

A practical approach with multiple authors is to nominate a lead author, who is usually listed first in an academic paper. For some publications, a similar role is as editor or a volume editor (Ritter, 2002). The lead author provides the leadership and the brief or plan upfront. This should include agreeing the template (or specific format requirements) and structure, who is writing which section and any guidance, such as a style guide. The lead author also feels ownership and is responsible for checking the whole document, in particular for any inconsistencies, duplications and omissions.

3.5. Organisation of ideas

It can be a challenge to synthesise complex, technical information and present it so that readers understand it. Information also needs to be managed, both for confidentiality and placement within a document, such as the main body or an appendix (see Section 2.4). The objective should be to convey technical information clearly and eliminate 'information dumping' in reports and other documents.

> Any message must be carefully organized before it can be either spoken well or written well.
>
> (Gunning, 1968)

A good structure with headings and sub-headings is the main way to organise ideas in a document and helps a document to flow (see Section 2.8). Structure varies with the rationale and purpose of a document. Where there is not already a template, developing a structure goes alongside organising ideas. There are several ways that ideas can be organised.

The next level down below sub-headings is to flesh out the skeleton by jotting down bullet points under each heading, so that the order of paragraphs flows within each section (see the start of Chapter 3). For example, a method section needs to cover the sequence of steps, the test sample and the test equipment. A logical flow of the bullet points might be from general to specific

- test equipment
- test sample
- sequence of steps.

A common order used is chronological, typical for many reports, where work is described in a time order, such as methods, results and conclusions. Another way to organise ideas is by the structure of a product or a project, for example, describing different work packages in a project.

How a document is to be used also affects the organisation of information. For example, engineering processes often start with prerequisites, such as tools required, any personal protective equipment (PPE) and safety warnings. This makes sense as some preparations may be needed before starting the task. The sequence of activities will then be step-by-step in order.

There are also standards for different types of document that provide structure and organisation of ideas. For example, software architecture or systems may be defined in nested layers or have software dependencies. Sub-system requirements are not needed if the system that they are part of is no longer required.

Another example of a standard for organising ideas is for specifications. For example, there is the British Standards Institution (BSI) standard on Product Specifications (BSI, 2001). Parts 1 and 2 of this standard cover products, and Part 3 covers services. The annex in Part 2 gives an example of a product specification template with headings, sub-headings and criteria under each.

Part 1 lists typical elements of a specification, with attributes under each one

- performance (function, material, dependability)
- location (environment, ergonomics and aesthetics, interface)
- acquisition and operation (cost and timing, training, safety)
- quality assurance
- legislation, health and safety.

The '5Ws and 1H' tool is a versatile device for many aspects of writing. This stands for

- what
- why
- where
- when
- who
- how.

Journalists use the 5Ws and 1H as a framework for a story in a newspaper article. It can be used for a whole document or just a section, such as an introduction to a report. For engineering documents, it is useful to describe a situation or define a problem statement, such as in an investigation or fault report. Some elements may be useful for a work package template (what, who, when, how much). There are variations on this tool used in problem solving, with additional questions, such as: how much? how many? how big?

It is useful to clarify a complete definition or description of a problem, particularly when it is an intermittent fault. Take care not to assume a root cause in a problem statement (the why). In the following example, saying that a leak is in the basement means that it is unlikely to be a leaky roof. However, knowing that it only happens when it rains might narrow it down to a leaky internal drainpipe.

- What? There is a pool of water on the floor.
- Why? Not known.
- Where? In the basement.
- When? It rains.
- Who? The security staff reported it.
- How long? For a week, intermittently.

'Bottom line up front' (BLUF) is a good way to communicate a message. As the name suggests, it starts with the bottom line, the key message. Rather than slowly building up to a recommendation, that is stated first, then substantiated.

This approach is good for grabbing attention and to save time for busy people. Engineers may have a natural tendency to want to explain the background fully before getting to the point, but a BLUF approach is sometimes better. Note that a summary contains more information and is different from a BLUF.

A similar concept to BLUF is an inverted pyramid, and there are variations in how this is applied. The most important information (or the answer) is presented first. This may be an opening paragraph, and the 5Ws and 1H may also be used to good effect here. Supporting explanations follow, then finally the details. Even if a reader does not read the entire document, they will still get the important ideas at the start. For those who need the detail to be fully convinced, they can read on.

An emphatic order is useful for presenting reasons or arguments, particularly in a persuasive document like a proposal. It may be used at different levels, such as paragraphs or a list of items. A logical order is better than a random one. The most important point is addressed first, then the others, reducing in importance. An emphatic order is sometimes used in the reverse order, such as when there are several strong reasons and the writer builds up to the clincher. The reader is most likely to remember the first or last items in a list, but less likely to recall any in the middle. Be careful not to put the most important point in the middle of a list, as it might get missed.

A spatial order may be used for descriptions of locations, such as from left to right, or from top to bottom. It can be useful to help readers visualise something.

Lateral communication refers to the same level in an organisation (rather than vertically up or down), typically information sharing among colleagues. This helps with collaborative working, sharing knowledge and exchanging ideas. While this is important, it tends to be less formal communication, such as using online platforms. As for other software tools, these need to be configured and used in a consistent way.

Summary of ways to organise ideas

- main body and appendix
- use a template with structure
- general to specific
- chronological order
- project structure
- process; start with prerequisites then step-by-step
- follow a relevant standard
- 5Ws and 1H
- BLUF
- pyramid
- emphatic order
- spatial order
- lateral communication.

3.6. Permissions and plagiarism

> Members must not commit or aid any form of plagiarism. Members must disclose the level of their use of AI or similar technologies [...]
>
> *Plagiarism*. Using the work of others without permission, acknowledgement or reference and presenting it as if it were your original work.
>
> ICE Code of Professional Conduct (ICE, 2026)

Plagiarism is an infringement of someone else's copyright or other IP, by using their material without permission. There is strong awareness of this in schools, colleges and universities, as the focus is on students producing original work, not copying others.

In academic study and research, it is expected that there will be a substantial literature review and a number of articles and other references cited. Work may be quoted, providing the use is fair and reasonable and within copyright laws, and is attributed to the author with full reference details. If the reference is paraphrased, then quotation marks are not required, but it is still necessary to give the source reference.

Schools and universities will use software to check for plagiarism. Experienced lecturers will often spot plagiarism as they are familiar with a body of research. A teacher shared an example with me, when she had set her class homework. Several students appeared to have done minimal research as they all used the same Wikipedia article, changing a few words here and there.

Photographs and other figures are also copyright information and may not be reproduced without permission. There is a difference between being able to do something and being allowed to do it. Some people do not seem to understand that because they can technically download an image from a website, it does not mean that they have permission to do so.

In publishing, formal written permission is usually required to reproduce someone else's IP, typically with a pro forma permission letter. There may also be copyright or permission issues when an author is presenting their own previously published work. Many journals only want to publish original work, not something that has already been printed elsewhere.

This topic is also linked to other aspects of writing, including protecting a company's IP (see Section 2.4), AI (see Section 2.7), online sources (Section 3.7), ethical writing (see Section 3.8) and referencing (see Section 4.10). Further detailed guidance is given in style guides (APA, 2020; Ritter, 2002) and the UK government's website (IPO, 2023).

3.7. Identifying reliable online sources

It can be a challenge to identify reliable online sources of information. Many websites have blogs, which are opinion pieces, without any references or peer reviews. Wikipedia is a great resource as an introduction to a subject, but it is not peer-reviewed and cannot be considered a reliable reference source. It even describes itself as 'the free encyclopedia that anyone can edit'.

Journal articles are considered the most reliable source of research. These are peer-reviewed and meet stringent criteria. Most engineering journals are published online now. Many published books are also available online as e-books. Self-published books are not as reliable a source, as they are not subject to the same rigorous editing processes. Google Scholar provides many articles and e-books, some of which are available in pdf file format, but others only through institutional access.

Many engineers will be able to access journals and e-books through their institutions. For example, professional bodies like the IMechE have a library. In addition to the physical library with books, they provide access to many online journals and e-books. Universities will similarly have institutional access to a wide range of journals. Companies may also have access to certain references by subscription, for example, to British Standards.

I will share an example I came across in researching this book. I wanted to find some research and statistics to quantify how much time was lost through distractions (see Section 3.1). I found a blog saying something along the lines of, 'It takes an average of 23 minutes and 15 seconds to refocus after being distracted at work'. In fact, I found several blogs and webpages referring to the same statistic in similar words, also referencing the same study and researcher.

Delving deeper, I found the original article referenced. However, I was unable to find that fact in the paper referenced, or indeed in any original academic paper. There are papers on that topic and by that author, but they do not mention 23 minutes. I found one article quoting the author as saying that in an interview, which may be the case. However, as that was reported second hand, I did not consider that a reliable source. What was surprising was how many websites had quoted a statistic without checking it.

I have other, similar, examples of multiple blogs or online articles quoting a statistic without giving the source. There is an oft-quoted statistic that only 7% of communication comes from words. While there was once a study with that number, this does not substantiate it as a general finding.

In summary, the best online sources are e-books from publishers (not self-published) and peer-reviewed journal articles.

3.8. Competency and ethical writing

> If you can't explain it to a six-year-old, you don't understand it yourself.

This and similar quotes have been attributed to Einstein and others, including Richard Feynman (who was awarded the Nobel Prize in Physics). This book is focusing on the skill of writing, which engineers can learn. Underpinning that, it is essential to understand the technical subject. In professional engineering, this means demonstrating competence in a field.

It is often obvious when people try writing about something that they do not understand. Consider a student doing a literature review, who writes that author A says, then quotes a long paragraph, then author B says, and quotes another long paragraph, and so on. That does not demonstrate understanding, as they have not synthesised what any of those quotations mean.

In the UK, the Engineering Council sets the standard for professional engineers (see Chapter 1). This standard is known as UK-SPEC (Engineering Council, 2020), and applies to all engineering institutions and other bodies licensed to award Chartered Engineer (CEng), Incorporated Engineer (IEng) and Engineering Technician (EngTech) status.

> Each registration title requires demonstration of competence in five broad areas
>
> A. Knowledge and understanding
> B. Design, development and solving engineering problems
> C. Responsibility, management and leadership
> D. Communication and interpersonal skills
> E. Professional commitment.
>
> (Engineering Council, 2020).

Alongside technical and engineering competences (A and B), communication skills is part of (D), one of the five core competences for engineering professionals. Technical engineering writing is a key communication skill. This extract for competence D from UK-SPEC is for CEng, although those for EngTech and IEng are similar. Note how the words 'clearly' and 'effective communication' are repeated here.

> D. Communication and interpersonal skills
>
> Chartered Engineers shall demonstrate effective communication and interpersonal skills.
>
> This is the ability to work with others constructively, to explain ideas and proposals clearly and to discuss issues objectively and constructively.
>
> The applicant shall demonstrate that they
>
> 1. Communicate effectively with others, at all levels, in English
> 2. Clearly present and discuss proposals, justifications and conclusions
> 3. Demonstrate personal and social skills and awareness of diversity and inclusion issues
>
> (Engineering Council, 2020).

In addition to the technical and communication competences, UK-SPEC covers ethical principles under competence E, and all three categories of engineers and technicians are required to, 'Understand the ethical issues that may arise in their role and carry out their responsibilities in an ethical manner.' (Engineering Council, 2020). Competence E also covers a personal commitment to comply with relevant codes of conduct, understand the safety implications of their role, and understand the principles of sustainable development and continuing professional development (CPD).

As required by UK-SPEC, engineers should also follow their professional institution's code of conduct, such as IMechE or ICE (ICE, 2025; IMechE, 2021). An extract from an example is given here.

> CR1. All members shall act with care and competence
>
> - They shall maintain their professional knowledge and are to maintain a record of continuing professional development.
> - They shall only undertake professional services in areas of their current competence.
> - They shall disclose any relevant limitations of their competence to undertake professional work.
> - They shall comply with obligations for health and safety. [...]
>
> CR2. All members shall act with integrity and in a reliable and trustworthy manner
>
> - They shall not knowingly mislead nor allow others to be misled.
> - They shall respect confidentiality obligations, express or implied.
> - They shall present and review engineering evidence, theory and interpretation honestly, accurately and without bias.
> - They shall avoid deceptive acts and refrain from corrupt practices.
> - They shall reject bribery and [...] record gifts or hospitality received.
> - They shall declare conflicts of interest or loyalty. [...]
>
> (IMechE, 2021)

Alongside UK-SPEC, the Engineering Council and the Royal Academy of Engineering jointly published a Statement of Ethical Principles for the engineering profession (Engineering Council and Royal Academy of Engineering, 2026). This was updated in February 2026, expanding from four to five principles, and now includes AI. The five fundamental principles to guide engineers and technicians in their professional life are

- honesty and integrity
- responsibility to society
- accuracy and rigour
- leadership and communication
- respect for the future of technology, society and the environment.

What the Statement of Ethical Principles and codes of conduct also highlight is the need to be honest, not misleading and without bias. Respecting confidentiality is also covered under plagiarism (see Section 3.6) and infringing IPR (see Section 2.4). Engineers need to declare any interest; for example, some scientific research has been challenged because researchers or authors failed to declare their funding source.

Pulling this all together, engineers need to be competent in their technical field and in their communications and act in an ethical way. Hence, the need for technical writing style to be clear and accurate, and for information to be presented objectively with a balanced view, supported by facts.

3.9. Reviewing and approving: stages and roles

Formal documents are reviewed by at least one other person. Sometimes referred to as the 'four eyes' principle, this means that someone else, in addition to the writer, has checked it. This ensures that these important documents are correct and complete, meeting the needs of all the stakeholders. Formal approval is recorded and the document is saved.

The reviewing and approving process should not start only after the document has been written. Doing so may also put a reviewer in a difficult position, if they are presented with a document to approve that they have not been involved with. It should start at the beginning of the writing process, with defining the brief and agreeing a plan (see Section 2.10). Following a process that starts with planning is the most efficient and effective way of writing, producing a quality document with the least amount of effort.

The planning stage covers several issues (see Section 2.10). Roles should be identified, for which I am using the terms proof-reader, technical reviewer and approver here. A template should be used or developed, see Table 2.1 and the expanded template checklist in Table A.1. Style guidelines should have been agreed, including options selected, see Tables A.2 and A.3. At the end of the planning stage, there should be an agreed master plan, for which Table 2.2 provides a checklist.

The reviewing and approving process resumes after the writer has drafted, edited and proof-read their own work. See Figure 3.1 for the review and approval process.

Note that there is a difference between proof-reading and editing. The definition of proof-reading is, 'the act of reading (copy or printer's proofs) to detect and mark errors to be corrected' (Collins, 2023). The definition of editing is, 'the process of correcting and adapting a text, such as an article or book' (Collins, 2023).

Figure 3.1 Review and approval process

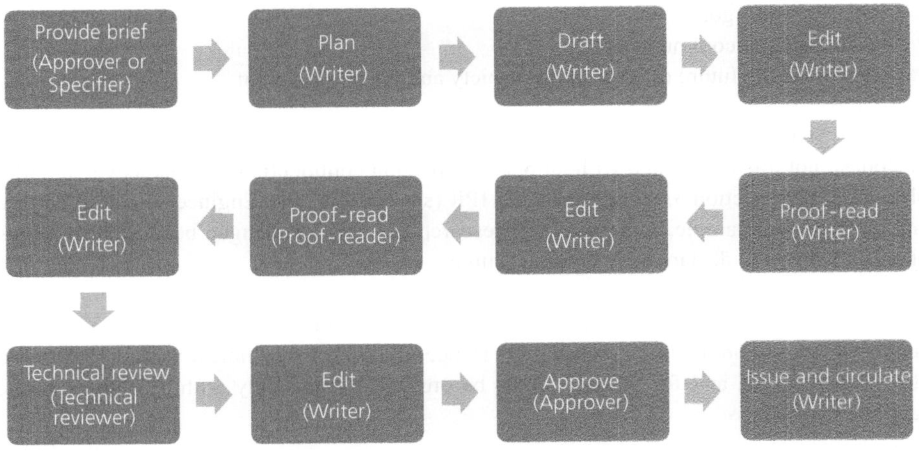

There are two things to note here. Firstly, that proof-reading is checking for errors, which implies an understanding of what is correct or not, and a checklist is useful to define this. Secondly, that proof-reading is only a check and does not extend to revising a document.

Only the writer should edit their document, and there are several reasons for this. First and foremost, it is the writer's document, and they need to keep that sense of ownership. If they lose that, it is demotivating and they can lose interest and personal investment in the quality of the output. They also need to understand the reasons for any corrections, so that they learn for the future, which may not be clear if others have made the changes. There may be a noticeable change in style if someone else edits it, which may mean that the writing does not flow well.

Nor is it the approver's job to make changes. Approvers are generally more senior people in an organisation and therefore their time is scarce and expensive compared with the writer's. I have heard of examples where senior people have modified a draft document without even telling the writer. Not only is that taking away the ownership, but there is also potentially a version control issue, and a greater risk of inconsistency. If an approver makes a change, then accepts it themselves, it also undermines the 'four eyes' principle.

There is an exception to this guidance, that only the writer should edit their own work, for some published documents, where there may be additional roles such as editor and typesetter (see Section 2.10).

There are typically, but not always, three people involved in the review of the completed document: the proof-reader, technical reviewer and approver. More steps may be added to the process in Figure 3.1 if needed. For example, if the approver makes any comments, then a further editing step would be needed. Sometimes there may be more than one technical reviewer, perhaps from a specialist function. It is important to follow this sequence in this order, even if individuals are asked to do more than one role.

Proof-readers are usually the writer's peers and familiar with the type of document. Unlike technical checkers, proof-readers do not have to be technical experts. For organisations that produce large numbers of documents, they may have a dedicated proof-reader role, who is not necessarily a technical person. They may not have that as a job title, in which case it is important to have a checklist, so that they (and anyone else) understand their role. The greatest load in reviewing is often in the proof-reading, as there are many points of detail to check for.

I have seen multiple examples of organisations with a 'checker' role that combines proof-reading and technical checks. Even if the same person is doing the proof-reading and technical checking roles, I recommend that they should only wear one 'hat' at a time, i.e. proof-read first, then do the technical check.

Unless the proof-reading changes are minimal, the writer should be allowed to make any proof-reading corrections before the technical check. Remember that SPaG changes may change the meaning.

Proof-readers are not usually part of the formal review and approval process, and their contribution is 'below the radar', not part of a formal procedure. Formal documents usually record the names of the reviewers and approvers as formal signatories but not proof-readers, which is as it should be.

I would strongly recommend that proof-reading is done before a document enters the formal review process. Where a process requires a record of any comments, these should not include any proof-reading issues, unless they have been identified after the document has entered the formal review process (and therefore the proof-reader has missed them).

A review form is typically used for formal communications under a contract with external third parties, such as customers and suppliers. Documents should be reviewed (and edited) internally first to minimise any such comments, as these may also be graded or classed and responding to every point can create a lot of work.

Some organisations require a formal trail of correspondence between the writer and the technical reviewer or approver, even for internal documents. This may be important for traceability, to check that a subject-matter expert's comments have been incorporated. However, unless a procedure requires a formal trail, their signature is usually sufficient to confirm that they are satisfied with the final version.

Tables A.4, A.5 and A.6 provide example checklists for the roles of the proof-reader, technical reviewer and approver, respectively. Note that the proof-reader has the longest list of issues to check for. For example, the proof-reader should check for details, such as spelling and labels; the technical reviewer should check technical content, such as equations; the approver should check that the document meets the agreed plan.

Another way to think of these roles is the primary focus within the four Cs (see Chapter 4): the proof-reader checks it is clear and consistent, the technical checker that it is correct, and the approver that it is complete.

I recommend that organisations develop their own version of these checklists, with the role names that they use and the points of details that they need to check for in their industry. Engineers may have different roles at different times. Even senior managers, who are mostly approvers, may be writers or checkers at times. As a writer, do not say to a colleague, 'Can you have a look at this?'; be clear which role is needed and what is to be checked.

It is good practice that both the writer and another person proof-read a document, reading it through fully. However, some of the proof-reading checks can be time-consuming, for example, ensuring that all acronyms and abbreviations are spelt out in full on first use and listed in the glossary. In this case, either the writer or the proof-reader (but not both) should check these items.

Similarly, only one person should check that the references are at the latest revision at the point of issue (or that there is a valid reason why a previous version should be used).

This might be considered a technical checker point (see Table A.5) rather than a proof-reader. For an effective use of time, either the writer or the technical checker (but not both) should check this point.

A proof-reader should check for points of detail such as homophones. These are words that are pronounced in the same way as another word, for example, 'weight' and 'wait'. Although they sound the same, they have very different meanings. Writers might hear the word in their head but write down the wrong one. As it is correctly spelt (although the wrong word), Word may not flag it. A common mistake is using the wrong 'their' or 'there'. I have seen 'the route cause', instead of 'the root cause'. Table A.7 lists some common and engineering-related homophones. As a fresh pair of eyes, a proof-reader is more likely to spot these.

Another homophone issue that I have seen is confusing prefixes and complete words, for example, 'non-' and 'none'. 'Non-' is a prefix that is attached to the start of another word (sometimes with a hyphen), for example, 'non-existent'; none is a stand-alone word, e.g. 'there was none'. This applies to other prefixes too. 'Sub-' is a prefix that can be applied to many words. The word 'sub' is sometimes used as a shorter form of 'submarine', 'subordinate' and other words. However, abbreviations should not be used in formal writing, particularly when they are ambiguous with multiple potential meanings.

I recommend that writers ask for feedback in person where possible, or at least live video, and be open to feedback that will improve their work. Even when there are marked-up documents or comment sheets, it is still better to hear directly from a reviewer or approver. In this way, the writer understands the concern and learns for the future. They can also explain, if there is a good reason, why something should not be changed. For example, a reviewer should not change the brief or request something be changed to their personal style.

Once company documents are formally approved, they should be stored in a document management system, so that they cannot be modified without creating a new version. It is good practice to retain previous versions if applicable, and any revisions should identify the reasons for the new issue. Unless procedures require, it is not normally required to retain copies of draft documents and any comments made.

Here are some tips on proof-reading.

- Use Microsoft Word to check SPaG.
- Use a checklist like Table A.4.
- Check one thing at a time for points of detail, e.g. figures labelled.
- Watch out for homophones.
- Print it out.
- Read out loud (or get Word to read aloud).
- 'Sleep on it'.
- Read it backwards.
- Get someone else to proof-read with a fresh pair of eyes.

REFERENCES

APA (American Psychological Association) (2020) *Publication Manual of the American Psychological Association: The Official Guide to APA Style*, 7th edn. APA, Washington, DC, USA.

BSI (2001) BS 7373: Product specifications. BSI, London, UK.

Collins (2023) *Collins English Dictionary*, 14th edn. Collins, Glasgow, UK.

Engineering Council (2020) The UK Standard for Professional Engineering Competence and Commitment (UK-SPEC), 4th edn. Engineering Council, London, UK. https://www.engc.org.uk/our-role-as-regulator/setting-standards/professional-engineering-competence-and-commitment (accessed 13/01/2026).

Engineering Council and Royal Academy of Engineering (2026) *Statement of Ethical Principles for the Engineering Profession*. Engineering Council and Royal Academy of Engineering, London, UK. https://www.engc.org.uk/media/4srbpyyu/statement-of-ethical-principles.pdf (accessed 04/05/2026).

Gunning R (1968) *The Technique of Clear Writing*, revised edn. McGraw-Hill, New York, NY, USA.

HMG (Her Majesty's Government) (1992) The Health and Safety (Display Screen Equipment) Regulations 1992. The Stationery Office, London, UK.

HSE (Health and Safety Executive) (2025) Working Safely with Display Screen Equipment. https://www.hse.gov.uk/msd/dse/ (accessed 04/05/2026).

ICE (Institution of Civil Engineers) (2026) *The ICE Code of Professional Conduct*. ICE, London, UK. https://www.ice.org.uk/media/usdi4cre/ice-code-of-professional-conduct-apr-26.pdf (accessed 07/06/2026).

IMechE (Institution of Mechanical Engineers) (2021) *Code of Conduct Regulations*. IMechE, London, UK. https://www.imeche.org/about-us/imeche-governance/governance-and-finance-reviews/code-of-conduct-explained (accessed 13/01/2026).

IPO (Intellectual Property Office) (2023) Intellectual Property: Copyright. https://www.gov.uk/government/collections/intellectual-property-copyright (accessed 12/03/2026).

Lindsell-Roberts S (2023) *Technical Writing for Dummies*, 2nd edn. Wiley, Hoboken, NJ, USA.

Ritter RM (ed.) (2002) *The Oxford Manual of Style: The Essential Handbook for All Writers, Editors, and Publishers*. BCA, London, UK.

Jean Billingsley
ISBN 978-1-83662-927-6
https://doi.org/10.1108/978-1-83662-924-520261005

Chapter 4
Writing style, English grammar, vocabulary

My four Cs of technical engineering writing are

- clear
- correct
- complete
- consistent.

For many years, I have delivered training following the principle of editing reports to address the 'three Cs' of writing, which are clear, concise and correct (Taylor, 2022). Others have similar guidelines, such as 'the ABC' of accuracy, brevity, clarity (Van Emden and Becker, 2018). Over time, in developing and delivering training, I have expanded this list to include other qualities, such as unambiguous, verifiable and traceable for engineering specifications (Billingsley and Taylor, 2023). I have also shifted from concise to complete, and added consistency for all engineering documents.

The overarching objective of technical writing is to be clear. This applies to many aspects of writing: structure, organisation of ideas, writing style and presentation. Technical writing must also be correct, with factual information presented in an objective way. While engineering documents should be concise, there needs to be a balance between being concise and being comprehensive, which is complete. Often the challenge is that a document or writing style is too long, but sometimes essential information is missing. Finally, any writing style, particularly for the same type of document, should be consistent. This gives a professional presentation and avoids ambiguity.

Good technical writing is about so much more than spelling, punctuation and grammar (SPaG) (which can be checked quite easily). Engineers should not get bogged down with grammatical terms, like fronted adverbials and digraphs. Engineers do not need to know these (and I am not going to explain them). It is not about being a stickler for the correctness of grammar for its own sake. Some joke about the 'grammar police' who just like to find fault. Even style experts argue against old perceived grammar 'rules' for more effective writing and less stiff sentences, such as ending a sentence with a preposition or splitting an infinitive (Strunk and White, 2000).

The point is that engineers should clearly communicate a message to the reader. This chapter explores some of the issues of choice of words, such as plain English. It highlights some points of detail to help avoid ambiguity, incomplete expressions or misleading associations between ideas. The use of grammar terms is limited to explaining concepts that can help with a clear and concise writing style, such as active sentences. A consistent approach also helps avoid ambiguity and provides a professional presentation. The objective of technical writing is to be clear, correct, complete and consistent.

4.1. Getting the message clear

As a writer, we want our reader to understand what we want to communicate. They should not feel ignorant or ill-informed because they do not know what a particular word means, do not follow an explanation or just generally find a document difficult to read. Too often, people reading badly written documents think that the problem is themselves, that they are not familiar with something that they should be. The onus is on the writer to communicate effectively – and to their target audience.

As a trainer, I sometimes do a fun quiz as a way to recap and test understanding. In one course on specifications, I covered the difference between verification and validation. In the first version, I found that quite a few delegates were getting that multiple choice question wrong. In reflecting on it, I challenged myself how I could better explain the difference so that delegates could understand the concept. Rather than just use words, I added a picture and thereafter found most delegates got that question right, so it had stuck. The point here is that, as a trainer, I took the responsibility to explain something in a way that people could understand and remember. Similarly, the onus is on the writer to write in such a way that the reader understands.

> Hi Jean,
>
> A few days ago, we sent you the wrong email – our bad! Sorry for the mix-up. Here's what we meant to say instead…

While most engineering documents should be formal, they should be clear first and foremost, and not excessively formal (see Section 5.10). A formal writing style is impersonal and assumes no relationship, and so avoids the first person (I and we) and third person (e.g. he) pronouns. It also avoids informal or colloquial language, such as 'our bad'.

4.2. Vocabulary and choice of words

It is important to choose the right words, not only in technical terminology but in everyday words too. Use a dictionary to check the correct and precise usage of words. A thesaurus can be useful for finding alternative words with a similar meaning, which might be a shorter word or simply a better word in the context. Using a variety of vocabulary is also less boring and more interesting for a reader.

I personally like to have a comprehensive dictionary and thesaurus by my desk, and I use online dictionaries and the thesaurus in Microsoft Word too. There are other resources available to help find the right choice of words.

One handy guide (Manser, 2014) has a subject index listed by heading, such as bids, essays and reports. Under each of these headings, it lists appropriate positive and effective words. For example, 'bid' includes advantage, budget, innovation and success. 'Essays' includes analyse, condition, reason and theoretical.

Avoid words like 'critical' and 'catastrophic', which may have specific defined meanings in a safety context. As technical writing should be factual and objective, avoid other exaggerated or hyperbolic language, such as 'spectacular' or 'alarming'.

Always check that technical terminology is used correctly too. For example, some people may use the words 'cement' and 'concrete' loosely as interchangeable terms. Cement is a binding powder, just one component of concrete; cement is also used in mortar.

Use words of more than two syllables sparingly (see Section 5.1). Sentences can often be rearranged with the same root words to be simpler.

- The complexity of this problem is challenging.
- This complex problem is a challenge.

Although shorter sentences are often clearer and easier to read, that is not always the case.

- Confidentiality in treating this matter is expected.
- It is expected that this matter will be treated in confidence.

Sometimes, a slightly longer sentence using simpler words is better. In the example above, the word 'confidentiality' is seven syllables and so more complex than 'confidence', which is only three syllables. The second sentence is also active, rather than passive (see Section 4.7) and so is clearer and more direct. This is not about chasing numbers to get the shortest sentence, as clarity is the most important thing.

Some people think that because it is an engineering document, this means that they have to use a lot of technical terminology. While it is right to use the correct and precise terms, there is still scope in the way it is written to keep it simple. For example, avoid excessive repetition of the same term several times within a paragraph, and particularly within a sentence.

Whilst this tip applies particularly to technical terms, it is also a good general guideline not to repeat words within a sentence or in consecutive ones. This can be avoided by using pronouns (such as 'it' or 'this') or rephrasing to simplify the language.

- There was significant structural deformation. This deformation was noted on the first unit, and the deformation was found to be due to high stresses.
- There was significant structural deformation. Noted on the first unit, this was found to be due to high stresses.

Consider these related words (Collins, 2023)

- Transcribe: 'to write, type, or print out fully from speech, notes, etc.'
- Transcript: 'a written, typed, or printed copy or manuscript made by transcribing.'
- Transcription: '1. the act or an instance of transcribing or the state of being transcribed. 2 something transcribed.'

Transcribe is a verb. Both transcript and transcription are nouns. I have seen people describe as a transcription something that I would consider is a transcript. It was a 'transcription' of a video recording. While transcription does mean transcript and can be used in that way, why would someone use a longer word with more syllables when there is a shorter word (from which it is derived) that says exactly the same thing? Unless referring to the act of transcribing, I would recommend using the word transcript for the end result.

Some words can even have two opposite or almost opposite meanings. There is even a word to describe this type of word – a contronym. For example, 'scan' is defined as, '1. to scrutinise minutely. 2. to glance over quickly,' (Collins, 2023). 'Act up' might be good or bad behaviour – deputising for a senior colleague or misbehaving. Be careful in the use of such words, as they may be ambiguous.

The word 'flammable' means exactly the same as 'inflammable'. Flammable means, 'liable to catch fire; readily combustible; inflammable' (Collins, 2023). 'Inflammable' means, 'liable to catch fire; flammable' (Collins, 2023). The true negative of flammable is non-flammable, 'incapable of burning or not easily set on fire; not flammable' (Collins, 2023).

A 'parking ticket' can be the slip of paper that someone pays for at a machine that gives them permission to park somewhere, or it could be the document that a parking attendant or parking enforcement officer has put on their car because they have parked where they should not have or longer than they should have for too long. The statement, 'I don't buy that,' could be taken at face value, that someone does not purchase that item. It could also be understood as meaning that they do not agree with something, with buy also defined as, 'to accept as true' (Collins, 2023).

The English language is a living language and the meaning of words can change over time, particularly with colloquial or slang usage. The term 'podcast' was only used for the first time in 2004. By 2026, podcasts have become very common. They are now sometimes abbreviated to 'pod', a term also used for a small herd of whales.

It is a skill to be able to communicate well at all levels, to technical peers and to lay people. This includes being able to explain technically complex ideas and what they mean. Some people think that because they are writing a formal, engineering document that it must be in very technical language. However, it is not a question of sounding clever. While it might seem counterintuitive, research has shown that when people write clearly they are perceived as more intelligent.

> We can conclude one thing. The pundits are likely right: write clearly and simply if you can, and you'll be more likely to be thought of as intelligent.
>
> (Oppenheimer, 2006)

4.3. Plain English

Plain English often means using a short and simpler word or using a short phrase rather than a long one. However, longer phrases may be better plain English if the words are simpler.

For plain English, one book (Cutts, 2013) has several pages of 'instead of ... consider using' words and phrases. For example, a suggested alternative to 'cognizant of' is 'aware of, know about'. There are also useful guides available from the Plain English Campaign (2025), including 'How to write in plain English' and 'The A to Z of alternative words'. For fun, their website even has a 'Gobbledygook generator', which produces 'a random piece of business jargon'.

I received an email, replying to one of my own, starting with the following:

> Hi Jean,
>
> Many thanks for your email below. I will pick up your points seriatim.

I had to look up what 'seriatim' meant in my dictionary. It derives from the Latin for series, and means, 'in a series; one after another in regular order' (Collins, 2023). While the word does have a very precise meaning, and was used correctly, it is not a well-known or commonly used word. It would perhaps have been better to say, 'point by point'.

When I am delivering a training course, I always ask delegates what they would like to get out of it. Someone said, 'maintaining concision'. The dictionary definition of 'concision' is, 'the quality of being concise; brevity; terseness' (Collins, 2023). While that word precisely defined what they wanted, the word 'concision' itself would not be familiar to most people. It would have been briefer and more understandable to say 'being concise'. We do not want our reader to feel ignorant or ill-informed because they do not know what concision, seriatim or some other word means.

I heard someone on the radio say the word 'complexifying'. This word derives from the word 'complexify', which means, 'to make or become complex' (Collins, 2023). According to the *Oxford English Dictionary* (OED) online, there are fewer than 0.01 occurrences per million words in modern written English. That means that word is used fewer times than one in a hundred million words. In this case, it would have been much clearer and better understood to say, 'to make complex'. For comparison, the word 'engineering' occurs about 20 times per million words.

The OED derives the frequency of use of words in written English and expresses each word (including its various forms and spellings) as occurrences per million words. Interestingly, there are only 5.2% of entries in the OED that occur with a frequency greater than one per million words. However, given that the dictionary has over 500 000 words and phrases, 5.2% is still over 26 000 words. Native speakers typically know 15 000 to 20 000 words in their first language (Sagar-Fenton and McNeill, 2018). I would suggest avoiding words with occurrences less than 1 per million words, and only using them if needed for a very specific technical term.

As for so many aspects of writing, the choice of words depends on the target reader and the context. The following two sentences from different style guides are making the same point (see Section 5.3 for this topic). The first (APA, 2020) is aimed at a more specialist publication audience and uses precise words relating to words and numbers, such as 'connote' and 'ordinal'. The second (Atherton, 2022) is aimed at a more general audience. (See Section 5.3 for this topic).

> The use of a numbered list may connote an unwanted or unwarranted ordinal position (e.g. chronology, importance, priority) among the items.
>
> If you use numbers (or letters for that matter) many, maybe most, readers will infer a hierarchy which they would not do with bullet points.

There is a greater risk of the wrong word being used with more complex words. Two examples I have seen of written documents or people using the wrong word are 'compromise' instead of 'comprise', and 'duplicity' instead of 'duplication'. Although the word compromise follows the word comprise in my dictionary, these words have very different meanings. I have also seen people confuse 'evasive' with 'invasive', 'pacific' and 'specific', 'alternate' and 'alternative', and 'comprehensive' (complete) and 'comprehensible' (understandable).

As a general guideline, do not use 'etc', 'e.g.' and 'i.e.' in formal documents (Collins, 2023).

- et cetera or etcetera, 'and the rest; and others; and so forth', abbreviated to etc.
- exempli gratia, 'for the sake of example', abbreviated to e.g.
- id est, 'that is (to say); in other words', abbreviated to i.e.

4.4. Writing for an international readership

This section focuses on writing for an international readership and non-native English speakers. It highlights some of the differences in US English and UK English and some useful resources. This book is aimed primarily at native English speakers.

There are many excellent references that can help people learn English. A series of self-study guides that I have used when working with non-native English-speaking engineers is the Cambridge 'In Use' series, which has exercises to practice each point of grammar (Hewings, 2013; Murphy, 2015, 2019). These books are aligned to the six levels of the CEFR (Common European Framework of Reference for Languages).

I often find that non-native English speakers have a better understanding of English grammar than native speakers as they have studied grammar as part of learning the language. I personally learnt more about English grammar from studying foreign languages than from English classes.

There are many differences between UK and US English, not only in spelling but also in punctuation, grammar and vocabulary. US spellings are often simplified compared with UK English: modeling vs modelling, color vs colour, aluminum vs aluminium and traveled vs travelled.

The letter 'z' (not 's') is more commonly used in US English; for example, 'analyze' not 'analyse'. In the USA, the spelling 'organize' is used; in the UK, either 'organize' or 'organise' is acceptable (Collins, 2023). Although the difference in spelling is unlikely to lead to misunderstanding, it can be a distraction. I have seen people with very strong feelings about not using the US spelling variants with a 'z', which risks detracting from checking more important content. There are differences in US English and UK English punctuation, even in what things are called. For example, period (US) or full stop (UK). The Oxford comma is used more in the USA, despite its name.

American style might give, 'I wrote you,' compared with British English, 'I wrote to you,' and 'I am good,' rather than, 'I am fine.' Grammar pedants may say that it is not grammatically correct to write these US examples. In the first one, it is missing the preposition 'to', as I did not write 'you' as an object. Some may consider the use of the word 'good' as an adverb in US English as colloquial; good as an adjective has many other meanings. See Section 4.5 on sentence structure for an explanation of word classes.

As for any point of style, consistency is key; adopt a preference and use it consistently. The proofing language can be selected for a template, so if a writer's client or parent company is in the USA, they might make a conscious decision to work with US English.

Another point to be aware of with an international reader is that guidelines for using capital letters may vary from one language to another. For example, in Spanish, languages or days of the week are not capitalised (as they are in English).

- Mi clase de español es el jueves.
- My Spanish class is on Thursday.

In English, capitalise

- the first word of a sentence
- proper nouns: names of people or places
- festivities, e.g. Christmas.

A useful resource when writing for an international readership is the Simplified Technical English standard (ASD, 2025). This is in two parts. The first has a set of writing rules, covering grammar and style and procedural writing. The second part is a controlled dictionary. This identifies a number of English words that have multiple meanings and gives a recommended usage. For example, the word 'replace' could mean 'put something back' or it could mean 'remove and substitute with a new one'.

4.5. Sentence and paragraph construction

Sentences are built from words, which are categorised as 'word classes', see Table 4.1. Many words can be multiple classes of word; a noun and verb combination is very common.

- Engineer (noun): 'a person qualified in a branch of engineering, especially as a professional'.
- Engineer (verb): 'arrange, contrive, or bring about, especially artfully, [...] act as an engineer'.

Table 4.1 Word classes

Word class	Explanation	Example
Noun	Person, place or thing	Jean, London, bridge
Verb	Doing or action word	Build, work, write
Adjective	Modifies a noun	Large, strong, smooth
Adverb	Modifies a verb or other word class	Regularly, badly, often
Pronoun	Substitute for a noun	They, it, some
Conjunction	Joins word, phrases and clauses	And, but, because
Preposition	Describes relationship to a noun	Over, by, across
Determiner (including articles)	Introduces a noun	The, many, both
Interjection	Exclamation	Whoa, hey, wow

It is best to use words in the class that they are commonly used for to avoid distractions. The phrase 'how to golf' is using the word 'golf' as a verb. While some dictionaries allow the word 'golf' as a noun or a verb, the most common usage is as a noun. Some readers may think there is a word missing and therefore get an impression that the document has not been fully checked. Others may argue that golf should not be a verb, and that is not the correct usage. Either way, if in doubt, it is best to avoid confusing or distracting the reader.

Conjunctions are joining words used within a sentence to link words, phrases or clauses. FANBOYS is an acronym to describe the seven common coordinating conjunctions: for, and, nor, but, or, yet and so. There are other conjunctions, such as: although, as, because and since.

There are two types of conjunction: coordinating and subordinating. The difference is to do with whether the linking parts are of equal importance in the sentence or not. Choose the right conjunction to convey the correct meaning and association of ideas. Take care not to imply cause and effect unless it is correct to do so.

- The test rig broke down and the electricity tripped off.
- The test rig broke down because the electricity tripped off.

Some people might argue that there is a 'rule' that sentences must not start with a conjunction, particularly in a formal writing style. However, this is not a hard and fast rule, and the practice may be used for stylistic effect. Allowing this element of style is one of the choices in Table A.3. Sentences should still be grammatically complete and the association of ideas clear (see Section 4.8).

Adjectives and adverbs are the two describing classes of word. For most formal engineering documents, these describing words should be used sparingly and avoided altogether for specifications as they are subjective. Adjectives and adverbs may be used in covering letters, for example, 'thank you for your kind words' or 'prompt response', and tend to be used more in persuasive writing than technical documents. Adverbs, such as 'then' and 'thus', may be used as linking or connecting words for follow-on sentences.

Linking words also help sentences and paragraphs to flow. These linking words or phrases are also referred to a transition words. They are often, but not always, at the beginning of a new sentence and followed by a comma where appropriate. They can serve several purposes. Table 4.2 gives some typical categories with examples.

Table 4.2 Transition words and phrases

Transition word category	Examples
Addition or list	Firstly, second, also, in addition to, further
Agreement or similarity	Similarly, likewise, equally, too
Introducing examples	For example, for instance, such as, particularly
Contrast	However, despite, by contrast, although
Condition	If … then, as, provided that, whenever, unless
Effect or result	Therefore, hence, as a result, for this reason
Location	Opposite to, under, alongside, where
Time or sequence	Next, then, after, once
Conclude or summarise	In conclusion, in summary, as can be seen, in either case

A document with a good flow is easy to read and helps the reader follow along and keep engaged. Flow is created at multiple levels; the structure (see Section 2.8), the organisation of ideas (see Section 3.5) and also in the writing style, which should be consistent throughout the document (see Section 3.4).

Maintaining the same tense, generally the simple past or the simple present, within a section helps writing to flow.

A nominalisation (or nominalization) is defined as, 'the conversion of a word or phrase into a noun' (Collins, 2023). Using verbs tends to be shorter and more dynamic, for example, clarify (verb) compared with clarification (noun). Some describe this as using strong verbs compared with weak nouns. The corresponding noun is often (but not always) longer than the verb. Also, as a sentence must contain a verb to be grammatically correct, another verb is required to complete a sentence.

- Provide clarification of the key points (nominalisation).
- Clarify the key points.

These verbs often add no value, only extra words. 'To carry out something' is a common example of nominalisation.

- The team carried out an inspection of the bridge (nominalisation).
- The team inspected the bridge.

Nominalisations are not grammatically incorrect, but are often wordy expressions and may sound excessively formal. Clear technical writing should avoid nominalisations.

> Verbalize or verbalise […] to change (any word that is not a verb) into a verb or derive a verb from (any word that is not a verb).
>
> (Collins, 2023).

A similar concept is verbalising, which is making a noun or other word into a verb, for example, to write 'to solution' is making the noun 'solution' into a verb. Another example is, 'productise your services'. Verbalisation is not grammatically incorrect, but it may mean using a longer word and, as not the common usage of the word, may not be as easily understood.

Anthropomorphism is, 'the attribution of human form or behaviour to a deity, animal, etc.' (Collins, 2023). This is a stylistic device used in some non-technical writing. However, an inanimate object cannot have human emotions or act like a human. Technical engineering writing style should be factual and objective. This is not to say that objects and non-human animals cannot be attributed with any actions, just not those that suggest they think or feel. Apart from not being a technical writing style, it can sound strange or unnatural. Phrasing of sentences should avoid this construction where thinking or feeling is implied.

It is acceptable for 'a beam to carry a load' and for 'a figure to show'. However, it is not appropriate to write that 'this study concludes' or 'the theory concludes'. Do not write, 'this report concludes that…' A report is an inanimate object and cannot conclude anything. The passive voice, in the shorter form (see Section 4.7), would be better here, 'it is concluded that…' Similarly, avoid, 'the building experienced an earthquake'; try 'the building withstood an earthquake'.

See also readability statistics (see Section 5.1) for guidance on word and sentence length.

4.6. Use of grammar and punctuation

Poor use of punctuation, such as a missing comma, can present a very different meaning from that intended, as illustrated by popular and entertaining reads, such as *Have You Eaten Grandma* (Brandreth, 2018) and *Eats, Shoots & Leaves* (Truss, 2007). One would hope that someone meant, 'Have you eaten, Grandma?', and was concerned for the welfare of their family member rather than talking to a cannibal. Poor use of punctuation may not give the intended meaning or may be confusing. Punctuation helps writing to be clear, correct and consistent. This section highlights some of the common areas of confusion.

See Table A.8 for the main punctuation marks and their uses (Billingsley, 2017; Seely, 2013; Trask, 1997). Other symbols and special characters are also used in engineering, such as curly brackets (sometimes referred to as braces) {} in computer programming and mathematical operators, less than, more than and plus-minus signs, < > ±, and also Latin and Greek symbols and currencies.

There are generally accepted conventions for the use of punctuation, although not universally fixed rules. Similarly, there is no single agreed nomenclature for punctuation, such as 'full stop' (or 'full point') vs 'period', 'brackets' vs 'parentheses', and 'apostrophe' vs 'inverted comma'.

Some of these differences are of British English and American English (see Section 4.4). As for any aspect of style, the writer (and ideally their organisation) should agree a preferred default term and use that consistently.

One variation is quotation marks. Either single or double quotation marks may be used for quotations; both are perfectly acceptable. This is one of the elements of style options in Table A.3. Like any aspect of style, once chosen, the same approach should be used consistently. If there is a quote within a quote, then use the other one (single or double) to make a distinction. I generally use double quotation marks for quotations and single quotation marks for 'so-called' or 'not-my-idea' expressions (Seely, 2013). As always, the objective is to be consistent throughout the entire document and to follow an organisation's style guide if it has one.

Authors note: This books uses the Emerald publishing style of single quotation marks for quotes.

Quotation marks are only used for direct quotes, not if paraphrasing a quote.

- He said, 'Your report is late,' (direct quote, with quotation marks).
- He said that their report was late (indirect quote, no quotation marks).

A formal writing style does not use some punctuation marks at all and others only sparingly. For example, exclamations and strong interruptions (which are similar to making exclamations) are not formal writing style, so the exclamation mark and em dashes should not be used in formal documents. An ellipsis may be used to remove words from a quotation but not to trail off at the end.

- 'The building works started in 2020 … and finished in 2024,' (acceptable use of an ellipsis in formal writing).
- 'Once upon a time…' (not a formal writing style for an ellipsis to trail off).

Many keyboards and editors use the same glyph for an apostrophe (') and the single closing quotation mark. When using an apostrophe to denote missing numbers or letters at the start, such as, in '99, Microsoft Word may default to an opening quotation mark during typing but should automatically correct to an apostrophe. This may be a point to check for.

- hyphen -
- en dash –
- em dash —

The hyphen, en dash and em dash may appear similar but are not the same length and have different functions, see Table A.8. The names for the dashes derive from the lengths of the letters 'n' and 'm', so the em dash is longer. There are no spaces around a hyphen. Do not mix words and the en dash, either write 'from … to' or use the en dash.

- The company makes ultra-precision products (hyphen after prefix to modify word).
- The range was 4–20 mA (en dash for a range).
- The control system—an archaic system—often tripped the plant (em dash for a strong interruption).

Microsoft Word should automatically change a hyphen to an en dash if spaces are added (type a number, then spaces before and after the hyphen, then type another number or word). Typing two hyphens in the same way will also give an en dash. To add an em dash, type a word, two hyphens and then another word (all without spaces) and Word should change it to an em dash. Note that autocorrect settings and Microsoft Word versions may vary.

Dashes may be another detailed point to check when proof-reading. However, as noted above, it is not formal writing style to use the em dash for a strong interruption, so this is something to be aware of, but not to use very often. The hyphen, en dash and em dash can also be found in Microsoft Word's Symbols under special characters or from shortcuts.

In many forms of writing, where text is justified, hyphens are commonly used to split words over two lines. This makes the presentation look pretty with left and right alignment. However, this may be less clear and can make a document harder to read because it splits words. Writing in engineering is about clear communication, often of technically complex ideas. Style and presentation choices should focus on clarity and understanding, particularly for internal engineering documents.

Although using hyphens in this way reduces 'stretching' of text across the line when justifying, it still does not eliminate extra spacing between words, which may look like typographic errors. This extra spacing is more noticeable in a justified two-column format, which journal articles often use. As noted in Table A.1, suggested template settings are 'do not hyphenate' and 'align left' (see Section 5.9).

Author's note: This book is printed in the Emerald publishing style, which is to justify and use hyphens to split words over two lines.

Some other hyphens make writing clearer and easier to read, for example fifty-two and pre-ignition. These are using hyphens for compound words and with a prefix, particularly where there are two consecutive vowels. Hyphens also help to avoid ambiguity with modifiers and multiple words.

For example, 'three hour long calls' may be confusing, whereas 'three-hour-long calls' or 'three hour-long calls' would be clearer. 'A small order fee' might be ambiguous. It could be that there is a fee only for a small order, or a small fee applicable to all orders. A hyphen could make this clearer, for example, 'a small-order fee', or rewording, 'a small order fee applies to orders under £40'.

Apostrophes have two purposes: contraction or possession. A contraction is where a letter or letters have been missed out. Like the ampersand symbol '&' (except for some citations), contractions are not used in formal writing but may be used in other documents

- can't – contraction of cannot
- won't – contraction of will not
- isn't – contraction of is not.

Apostrophes are also used to denote possession, and the position depends on whether there is one or more subjects. If singular, the apostrophe goes before the letter ‘s’ added to the end of the word; if plural, the apostrophe goes after the ‘s’. If the subject is already a plural noun, it is treated as singular with the apostrophe before the ‘s’.

- The building’s drawings were electronic. (There is only one building.)
- The buildings’ drawings were electronic. (There is more than one building.)
- The women’s facilities were poor. (Plural noun, apostrophe before the ‘s’.)

A special case is for possessive pronouns, which are already in the possessive form so do not have an apostrophe added: its, his, hers, yours and theirs. Its means belonging to it; it’s means it is

- its (possessive pronoun, belong to it)
- it’s (contraction, meaning ‘it is’).

There is no universal agreement for the case of a subject that already ends in an ‘s’. Some add the apostrophe and an ‘s’ as for other singular subjects; others only add the apostrophe. As for other points, this could be covered in a style guide

- James’s PC (possessive of singular subject with extra ‘s’)
- James’ PC (possessive of singular subject without extra ‘s’).

A common error is an incorrect apostrophe for a plural, sometimes referred to as the ‘grocer’s apostrophe’ (because it is apparently often seen in grocery stores), where there is an apostrophe but there should not be, for example, a sign saying

- banana’s for sale (incorrect)
- bananas for sale (correct).

The comma is a versatile punctuation mark with several different purposes; the main ones being covered in Table A.8. However, commas should be used in moderation within one sentence, ideally not mixing different types of comma to avoid confusion. For example, brackets may be used instead of bracketing commas when there are other commas in the sentence.

The ‘Oxford comma’ is a particular type of listing comma, used before the final ‘and’ in a list. It is recommended by Oxford University Press (OUP) style guides (Ritter, 2002), hence the name. However, it is generally used more in US English and the convention in UK English is often not to use the Oxford comma. It may be used for clarity, for example if the final item is a long phrase or contains an ‘and’.

- Available colours were red, yellow, green and blue. (no Oxford comma and clear)
- Available colours were red, yellow, green, and blue. (with Oxford comma and clear)
- Human Resources were recruiting for Design, Operations and Research and Development. (no Oxford comma and unclear if ‘Research’ and ‘Development’ are one or two departments)
- Human Resources were recruiting for Design, Operations, and Research and Development. (with Oxford comma and clear that ‘Research and Development’ is one department)

Some punctuation issues can be addressed by settings in the template, for example 'do not hyphenate'. A lot of punctuation errors can be picked up by Microsoft Word checks and also proof-reading by the writer and the proof-reader.

A good general guideline is to add punctuation, such as a hyphen or the Oxford comma, where it helps to make it clearer. As with other aspects of writing, sometimes it is better to rewrite a sentence in another way to avoid confusion. Good use of punctuation makes writing clear and easy to read and can avoid ambiguity.

4.7. Style: active and passive

Many writing and style guides advise avoiding the passive. Let us start by understanding what we mean by passive, why some style guides do not recommend it and when it may be used.

Verbs are the class of word which are actions or doing words. As the words verbosity and verbiage suggest, the use of verbs can often contribute to wordiness or wordy habits. Verbose means, 'using or containing an excess of words, so as to be pedantic or boring' (Collins, 2023). Engineering documents should be complete, which means as concise as they can be without omitting key information or introducing ambiguity. We can often look at a sample of writing and be able to write the same thing more clearly and concisely.

Active and passive forms of a verb are sometimes referred to as the active and passive voices. In grammatical terms, the difference is the relationship between the subject and the action. In an active sentence, the subject of the sentence is doing the action. In a passive sentence, the subject is receiving the action or having something done to it.

- Cheshire East Council maintains roads. (active)
- The roads are maintained by Cheshire East Council. (passive)

In the first example, the subject of the sentence is Cheshire East Council, and they are doing the action of maintaining the roads. In the second example, the subject of the sentence has become the roads, although Cheshire East Council are still doing the maintenance. The roads are inert and are not doing anything.

As seen in this example, although the 'doer' is the same, the word order is different. In the passive sentence, 'by the doer' usually comes at the end of the sentence. This means that the reader does not find out who is doing the action until the very end of a passive sentence. By contrast, an active sentence starts with the 'doer' (who is also the subject) and is therefore clearer. Active sentences are also shorter, typically two or three words fewer than the related passive sentence. That might not sound like much but can be quite a significant percentage. Active sentences are generally preferred as they are more direct and engaging, in addition to being clear and concise (Grimond, 2013; Strunk and White, 2000).

Passive sentences have their uses. They can be used for emphasis, to focus on the results or the receiver rather than the person doing something. In the example above, it is putting the focus on the roads not the council.

Passive sentences can also help to avoid repetition when the doer is not relevant or important and the doer (or 'agent') may be omitted, which cannot be done in an active sentence. When an engineer is writing a report on their own work, then it is implied that they or their organisation did the work, and they do not need to start every sentence with I. In this case, the shorter form of a passive sentence should be used, omitting the doer. This shorter form is called the 'truncated passive' (as it has the end of the sentence cut off) or is sometimes referred to as the 'agentless passive', or 'passive with an unknown actor'.

- I recorded daily measurements. (active)
- Daily measurements were recorded by me. (passive)
- Daily measurements were recorded. (truncated passive)

This truncated form of the passive is also shorter than the full passive. It is usually about the same number of words as the active and so helps to keep writing concise. This form of the passive is also useful for writing formal documents as it avoids the use of personal pronouns, such as I or we (sometimes referred to as the 'first person'; see also Section 5.10 on house style). It is a common convention for a formal writing style to avoid using the first person.

However, it is more important to be complete than concise, and the writing should not be ambiguous or have any mystery. If it is important to know who has done the action, then the doer should be identified and the sentence written in the active form.

- Changes to the contract have been agreed. (truncated passive, unknown actor)
- The project manager agreed changes to the contract. (active with actor)
- Action needs to be taken. (truncated passive, unknown actor)
- Management needs to act. (active with actor)

In summary, active sentences are generally preferred as they are clear, concise and direct. Passive sentences may be used for emphasis or in formal documents where the doer (actor) is not important. See also Section 5.1, on readability statistics.

4.8. Ambiguity and other common problems

Not only do we want our readers to understand what we write, but we also need to make sure that they cannot misunderstand what we write. Ambiguity in writing can arise not only from poor grammar and punctuation (See Section 4.6) but also several other sources. These include vague words, words that have multiple meanings, similar sounding words (see Section 3.9 and Table A.7), the order of words and the association of ideas.

Many words have multiple meanings. To engineers, the word 'gear' means a toothed wheel. It also has several other uses, such as equipment (as in landing gear), informal use for up-to-date clothing or slang for illegal drugs.

Avoid using adjectives and adverbs to avoid ambiguity (see Section 4.5). What is a 'good' quality, and how long is 'quickly'?

There is a difference between the definite article 'the' and the indefinite article 'a' or 'an'. Using articles makes a clear distinction between something specific and something general, for example, 'the building' and 'a building'. Do not omit articles just to make a sentence shorter as this can be ambiguous. It might also make a difference between a requirement being achievable or not.

Some people may say that starting a sentence with 'and' (or another conjunction) is not grammatically correct (see Section 4.5). However, what we are focusing on here is writing to communicate clearly rather than blindly following 'rules'. Having said that, there is a risk that the reader may be distracted if what they perceive as a rule is broken. Sentences starting with a conjunction may also need more care in their construction. Conjunctions, such as 'and', are joining words, and are meant to join parts of a sentence together. If a sentence starts with a conjunction, then it may lose the correct association of the ideas.

Another potential cause of ambiguity, similar to losing the association of ideas by starting a sentence with a conjunction, is an incomplete expression.

The words 'if' and 'whether' are similar in some respects but have slightly different usage in others. The word 'if' implies a condition and may be used in the form of 'if … then'. Both 'if' and 'whether' may be used for an indirect question where the answer is yes or no (and optionally 'or not' may be added at the end of the sentence). 'Whether' introduces two options and may be used with 'or not' or another alternative. For completeness and to avoid ambiguity, I would recommend using 'or not' or stating an alternative when using 'whether'.

- If we win the tender, then we will need extra staff.
- They did not know if they would win the tender (or not).
- They did not know whether they would win the tender (or not).
- She did not know whether to recruit now or next month.

The word 'confirm' means, 'to prove to be true or valid; corroborate; verify' (Collins, 2023). 'If' means, 'in case that, or on condition that' (Collins, 2023), and might be ambiguous combined with the word 'confirm'.

For example, 'confirm that…' or 'advise if…' would be better than 'confirm if…':

- She asked them to confirm if they were going to the meeting. (incorrect)
- She asked them to confirm that they were going to the meeting. (correct)
- She asked them to advise if they were going to the meeting. (correct)

Other expressions need to be balanced and complete. These include: 'both … and', 'not only … but also', 'either … or' and 'neither … nor'. These need to be balanced and complete. The word 'either' requires 'or' to follow, and 'neither' requires 'nor'; this should be a two-part phrase and is otherwise unfinished.

The problem was neither the starter motor nor the battery.

Another word to use carefully is 'compared. 'Compared to' suggests resemblance; 'compared with' suggests difference (Strunk & White, 2000).

Similarly, the association of ideas is incomplete if the comparative or superlative forms are used in isolation. The context may be explained in the previous sentence, but should be there somewhere. The comparative means compared with, relative to, or to a higher degree, and is usually created either by adding 'er' to an adjective or using the word 'more', for example, 'greater' or 'more expensive'. The superlative means the highest or greatest degree of something. It is usually created by adding '-est' to an adjective, or using the word 'most', for example, 'greatest' or 'most expensive'.

- The larger site was selected (incomplete, compared with what?).

The word 'while' can be used in multiple ways. Take care not to write an incomplete sentence, such as, 'While the concrete is setting.'

Ambiguity can also be caused by the word order in a sentence (the grammatical term here is the 'syntax'). I once sent an email about an insurance claim writing,

> They [the insurance company] have sent the attached to the resident who reported the incident in error.

I meant that the insurance company had sent something to the wrong person. The resident concerned replied immediately to say that they did not report the incident in error! The following would have been better:

> They [the insurance company] have sent the attached in error to the resident who reported the incident.

In this case, the construction of the sentence with the word order is the problem.

Be careful with the word order also for the 'understood subject' (the grammatical term here is the 'antecedent'). This means making sure that it is clear who is doing the action. While pronouns may be used as a substitute for a noun to avoid repetition, take care to avoid ambiguity. Consider the following:

> The Supplier shall submit the drawings to the Client. They shall provide three hard copies of each drawing.

Who is going to provide the hard copies? As the client is the last person mentioned, then the common convention of the 'understood subject' would suggest the client, which might not be the author's intent. For specifications and to avoid any ambiguity, the safer choice would be to repeat the word 'supplier' rather than using the pronoun 'they'.

The words 'that' and 'which' can both introduce clauses. The difference is 'that' introduces a defining clause, something that is essential to identification and the meaning of the sentence. 'Which' introduces a non-defining relative clause, typically some extra information. There is no additional punctuation with 'that'; which is used with bracketing commas (see Table A.8). Even with similar words, the meaning conveyed is different.

- The report that he wrote was approved.
- The report, which he wrote, was approved.

Ambiguity summary

- Be careful with words that have multiple meanings.
- Generally avoid adjectives and adverbs.
- Do not omit articles.
- Do not start a sentence with a conjunction.
- Ensure expressions are always complete, e.g. whether or not.
- Be careful with the association of ideas.
- Ensure the word order gives the correct meaning.
- Be clear who the subject is, avoid confusion with the understood subject.

4.9. Building a glossary

A glossary is important for all the four Cs: for a document to be clear, correct, complete and consistent. There are often many acronyms, abbreviations and specific terms used in engineering documents.

An acronym is a word formed from the initial letters of other words. These may be things or organisations. For example, Institution of Civil Engineers (ICE), and automatic number plate recognition (ANPR). The same acronym may have multiple different meanings, which is one reason for defining them. RMC is an abbreviation for ready-mixed concrete; the same acronym can be for residents' management company. SME might be 'subject-matter expert' or 'small and medium-sized enterprises'. Some acronyms are pronounced as words, such as CAD and SQEP; others (defined as initialisms) are pronounced as separate letters, such as AI and PPE.

There are abbreviations that are not strictly acronyms as they have more than the first letter of a word. Examples include Institution of Mechanical Engineers (IMechE), Institution of Structural Engineers (IStructE) and Chartered Engineer (CEng). Other forms of abbreviation may also be used, for example, HMS PWLS is used in signals (Navy radio messages) for HMS *Prince of Wales*, and HMS *Queen Elizabeth* is HMS QNLZ.

Define acronyms and abbreviations in full the first time they are used followed by the abbreviation in brackets, and also put them in a glossary. This helps the reader, because if they dip into the middle of a document and find an acronym that they are not familiar with, they can look it up. If an abbreviation is only used once in a document, it is unnecessary to put it in the glossary. Nor is it necessary to do so if the abbreviation is well known in the context, such as UK. Note that this book lists all acronyms and abbreviations even if only used once as that is the Emerald publishing style. A glossary may be placed at the beginning or the end of a document, outside of the main body. Many formal engineering documents will have a glossary.

A glossary may also describe words that are given a specific meaning in a contract, specification or similar document, in which case the word in the body of the text has a capital letter to indicate that it is defined, for example, the Authorised Person, the Engineer or the Contractor. Where terms are titles or proper nouns, they should be defined with capital letters. If they are an abbreviation for a common term, then lower case letters may be used.

Do not use acronyms and abbreviations in headings or titles (and ideally not in references or a summary either). While acronyms are useful, they should be used in moderation. While organisations do have acronyms and abbreviations specific to their products and services, take care not to create new ones unless they are likely to be used more widely. Readers will be unlikely to know them. Where the acronym or abbreviation includes a word, it should not be duplicated. For example, PAT stands for 'portable appliance testing', so writing 'PAT testing' is duplicating the word 'testing'.

After time, some acronyms get adopted as words. For example, the word laser was originally an acronym LASER (light amplification by stimulated emission of radiation); 'scuba' came from SCUBA (self-contained underwater breathing apparatus); 'sonar' was sound navigation and ranging, after radar, radio detection and ranging (Pearsall and Trumble, 1996). If in any doubt, spell it out.

4.10. Referencing

When using references, ensure that all the details are captured at the original point of use. Although these should be checked in the reviewing and editing process, that should not mean having to repeat work. This is another instance of integrating writing with work, recognizing that the reference source will be needed for the document. It can be difficult to find sources again at a later date and online content may change.

There is a difference between references and bibliography. References are specifically referred to in the body of the document. A bibliography is there for background reading, but is not specifically referred to. However, it is unusual to see both in an engineering document, and sometimes the word 'references' is used as a catch-all.

References may be presented in different formats. At the time of writing this book, Microsoft Word presents references in either 'APA 6th edition' or 'MLA 7th edition'. These refer to the American Psychological Association (APA), whose *Publication Manual* is currently at the 7th edition (APA, 2020), and the Modern Languages Association of America (MLA), whose handbook on style is currently at the 9th edition (MLA, 2021).

When quoting, write the quotation, then insert the citation and put the full stop at the end. Use an ellipsis (three full stops with no extra space) where words have been omitted. A quotation is defined as, 'a phrase or passage from a book, poem, play, etc. [...] an estimate of costs submitted by a contractor' (Collins, 2023).

REFERENCES

APA (American Psychological Association) (2020) *Publication Manual of the American Psychological Association: The Official Guide to APA Style*, 7th edn. APA, Washington, DC, USA.

ASD (Aerospace, Security and Defence Industries Association of Europe) (2025) *ASD-STE-100. Simplified Technical English: Standard for Technical Documentation*, 9th edn. ASD, Brussels, Belgium.

Atherton T (2022) *Technical Report Writing and Style Guide: How to Write Even Better Technical Reports*, 7th edn. Amazon, Seattle, WA, USA.

Billingsley J (2017) *So What? Effective Writing for Engineers*, 1st edn. CreateSpace, Scotts Valley, CA, USA.

Billingsley J and Taylor P (2023) *Preparing Engineering Specifications*. IMechE, London, UK.

Brandreth G (2018) *Have You Eaten Grandma? Or, the Life-Saving Importance of Correct Punctuation, Grammar and Good English*. Michael Joseph, London, UK.

Collins (2023) *Collins English Dictionary*, 14th edn. Collins, Glasgow, UK.

Cutts M (2013) *Oxford Guide to Plain English*, 4th edn. Oxford University Press, Oxford, UK.

Grimond J (ed.) (2013) *The Economist Style Guide: The Bestselling Guide to English Usage*, 10th edn. The Economist, London, UK.

Hewings M (2013) *Advanced Grammar in Use: A Self-study Reference and Practice Book for Advanced Learners of English*, 3rd edn. Cambridge University Press, Cambridge, UK.

Manser MH (2014) *1001 Words You Need to Know and Use: An A–Z of Effective Vocabulary*. Oxford University Press, Oxford, UK.

MLA (Modern Language Association of America) (2021) *MLA Handbook*, 9th edn. MLA, New York, NY, USA.

Murphy R (2015) *Essential Grammar in Use: A Self-study Reference and Practice Book for Elementary Learners of English*, 4th edn. Cambridge University Press, Cambridge, UK.

Murphy R (2019) *English Grammar in Use: A Self-study Reference and Practice Book for Intermediate Learners of English*, 5th edn. Cambridge University Press, Cambridge, UK.

Oppenheimer DM (2006) Consequences of erudite vernacular utilized irrespective of necessity: Problems with using long words needlessly. *Applied Cognitive Psychology* **20(2)**: 139–156, 10.1002/acp.1178.

Pearsall J and Trumble B (eds) (1996) *The Oxford English Reference Dictionary*, 2nd edn. Oxford University Press, Oxford, UK.

Plain English Campaign (2025) Free Guides. https://www.plainenglish.co.uk/free-guides (accessed 20/02/2026).

Ritter RM (ed.) (2002) *The Oxford Manual of Style: The Essential Handbook for All Writers, Editors, and Publishers*. BCA, London, UK.

Sagar-Fenton B and McNeill L (2018) How many words do you need to speak a language? https://www.bbc.co.uk/news/world-44569277 (accessed 11/05/2026).

Seely J (2013) *Oxford A–Z of Grammar & Punctuation*, revised 2nd edn. Oxford University Press, Oxford, UK.

Strunk W and White EB (2000) *The Elements of Style*, 4th edn. Pearson, Upper Saddle River, NJ, USA.

Taylor P (2022) *Technical Report Writing*. IMechE, London, UK.

Trask RL (1997) *The Penguin Guide to Punctuation*. Penguin, London, UK.

Truss L (2007) *Eats, Shoots & Leaves: The Zero Tolerance Approach to Punctuation*. Fourth Estate, London, UK.

Van Emden J and Becker L (2018) *Writing for Engineers*, 4th edn. Bloomsbury, London, UK.

Technical Engineering Writing: A Practical Guide for Professional Engineers

Jean Billingsley
ISBN 978-1-83662-927-6
https://doi.org/10.1108/978-1-83662-924-520261006

Chapter 5
Presentation

Presentation covers many aspects of writing clearly and consistently, including formatting of paragraphs and bullet points. This chapter also covers numbers, units, equations, figures and tables, making documents accessible, and the benefits of style guides. However well laid out documents are, they can still be very off-putting to the reader if they are not presented in a readable style and this is where readability statistics are a useful benchmarking tool.

5.1. Readability statistics

Sometimes when readers find something hard to read, they put it down to themselves, that they do not understand a subject well enough. However, it may be the case that the text is simply written in a way that is hard to read. Chapter 4 covers writing in a clear, correct, complete and consistent style. Building on that, readability statistics provide a measure of how readable a sample of writing is and how it can be improved.

In essence, these 'readability statistics' do exactly what they sound like – they are a measure of the readability of a sample of writing. This is an objective measure, not subjective feedback, and many engineers find these useful as they like to work with numbers. As a general guide, longer words and longer sentences are harder to read.

Readability 'yardsticks' have been around since the 1930s (Gunning, 1968), including the Gunning fog index, which was based on the number of words in sentences and the number of words with three or more syllables. This is similar to the readability statistics that Microsoft Word now generates automatically, with the added advantage of not having to manually count words in a sample.

The Flesch–Kincaid 'grade level' ('grade level' for short) and the Flesch 'reading ease' ('reading ease' for short) are both composite statistics derived from the same two inputs: average number of words per sentence and average number of syllables per word. They have slightly different weighting factors between the two inputs but broadly give similar information. One thing to note is that they are an inverse scale of each other: the higher the grade level, the harder it is to read; the higher the reading ease, the easier it is to read.

Like anything with statistics, a representative sample size is needed, ideally a couple of pages of writing in paragraphs. To generate these statistics in Microsoft Word, go to the Review tab, select Editor on the top left ribbon and the Editor pane will open on the right-hand side of the screen. Then, scroll down the Editor pane to Insights, and select Document stats. A table will appear similar to the example in Table 5.1.

For some reason, Microsoft Word's calculations of the averages do not always appear to be the correct mathematical average, consistent with the counts. The recommended targets in Table 5.2 are based on the readability and averages in the readability statistics output.

For previous versions of Microsoft Word that do not have the Editor, the readability statistics can still be found but with a few extra steps. From the Review tab, Language drop-down, Language preferences, select Proofing from the left-hand bar. Ensure that the check box for readability statistics is checked and save these settings by clicking on OK. Then complete a spelling check, and the statistics will appear. Note that the spelling check must be completed. If it is stopped, then no statistics are provided. If a document is very large and not yet checked, select a couple of pages and copy and paste them into a new document to check.

Table 5.1 Readability statistics example output

Readability statistics	
Counts	
Words	29102
Characters	155159
Paragraphs	1457
Sentences	1540
Averages	
Sentences per paragraph	2.3
Words per sentence	15.5
Characters per word	4.8
Readability	
Flesch reading ease	53.5
Flesch–Kincaid grade level	9.5
Passive sentences, %	15.1

Table 5.2 Readability statistics target

Statistic	**Target**
Flesch–Kincaid grade level	9–11
Flesch reading ease	>50
Average characters per word	≤5.0
Average words per sentence	15–20
Percentage passive, %	<30

Note that the recommended targets are based on writing in paragraphs. If a sample of writing has heavy use of lists, tables or references, then it may skew the results.

The counts are a useful check that the readability statistics are sensible numbers. If these are all zeros, or much lower than the size of the document might suggest, then it may not have worked correctly. This is typical in a heavily formatted document, in which case, select the sample and copy as text only to a new blank document and then try again. Another reason may be with compatibility. Check if the sample document has been saved in the latest version of Microsoft Word.

The readability numbers are in the lowest third of this table. A guideline target for technical writing is given in Table 5.2. These guidelines have been generated from research and have been borne out in practice in my experience. Note that these guidelines are based on characters per word, rather than syllables per word that the formulas use, but there is a correlation between these two.

These targets recognise that technical engineering writing is going to be harder to read than a typical novel and many other styles of writing. For example, many general style guides recommend avoiding the passive altogether. However, it has its uses (see Section 4.7).

For a technical report, Tony Atherton recommends targets of 17 to 22 words per sentence, 25% to 45% passive, and 45 to 60 for reading ease (Atherton, 2022). Robert Gunning describes hard-to-read sentences as over 20 words (Gunning, 1968). Some recommend an average of 15–20 words (Cutts, 2013; Seely, 2013), and others 17–20 words (Van Emden and Becker, 2018). The Web Content Accessibility Guidelines (W3C, 2024) recommend a reading level of lower secondary education level, which they define as nine years of schooling.

Focusing on the average number of words per sentence is useful, as some variety in sentence length helps the reader by making reading more interesting. Short sentences can be used for emphasis. Consider the impact of the short and direct sentence, 'The test failed.' Sentence length is something to be aware of, particularly when explaining technically complex ideas. Specifications and procedures, which state requirements and instructions, should be shorter, tighter sentences than general technical writing (see also Section 6.2).

Some give recommendations as to the maximum length of a sentence. For example, use a maximum of 20 words for a procedural sentence or 25 words for descriptive writing (ASD, 2025), or a maximum of 40 words (Van Emden and Becker, 2018).

It is interesting to note that early research and readability formulas looked at a 'hard-word factor' as those not on a list of 3000 familiar words. Only 4% of these 3000 words exceeded two syllables (Gunning, 1968). As there was a correlation between these 'hard words' and words with three or more syllables, formulas were developed around syllables. However, it is useful to keep in mind the derivation of these formulas, as it is the less well-known words that are harder to read, rather than longer words per se (see Section 4.2).

Making writing more readable is not 'dumbing down'. It is about clear communication so that the reader gets the intended message and acts on it. When trying to explain technically complex ideas, it is even more important to communicate clearly so that the reader understands. Remember, always focus on the reader. A document with a lower grade level is simply easier to read.

> Be sparing, therefore, in your use of hard words and marathon sentences, so your reader will have energy left for your complicated ideas.
>
> (Gunning, 1968)

Although Robert Gunning was an early pioneer of using readability yardsticks, he advocated principles rather than rules and formulas for clear writing. Keep in mind that these statistics are only guidelines, 'a tool not a rule'.

5.2. Formatting for ease of use

Keeping the reader in mind, clear presentation and ease of use helps them. Formatting techniques can enhance readability. Clearer presentation makes it easier to read and signposts help the reader navigate the document and find what they want. This includes a table of contents, paragraphs, bullet points and labelling figures (see Section 5.8). The good news is that saving settings in a Word template means that these points will be consistent whenever that template is used. This applies for all settings, including proofing language, font and format. See Section 2.8 and Table A.1 for a template checklist.

I would always recommend a table of contents unless a document is very short, perhaps fewer than ten pages. It is a useful signpost to help the reader find the section that they want. Number the sections and use an automatic table of contents with page numbers. In Microsoft Word, a table of contents can be added from the References tab. Avoid too many levels in a contents list, ideally only one or two levels and a maximum of three levels for a very large document (see Section 2.8). For most documents, the table of contents should fit on one side of A4 paper.

If the document has a large number of figures, then it can be useful to insert a table of figures or a table of tables too. As an extra point of navigation, it can provide further signposting for the reader.

The starts of paragraphs may or may not be indented. Some styles have all the opening sentences of paragraphs indented; others have none. Some styles (typically books) do not indent the start of the first paragraph in a section but then indent subsequent ones. Paragraphs are sometimes separated by a line space, particularly when their starts are not indented.

For most documents (although occasionally some books do this), the rest of the paragraph should not be indented, even for sub-sections, as it wastes a lot of the available space on the page. Indentations may be used for quotations and examples. Other recommended paragraph settings are widow and orphan control (to prevent a single word at the end of a paragraph or a single line on a separate page), and check the box to not hyphenate (see Section 4.6).

Some people use internal links within a document, such as cross-referencing captions. Links to external sources may also be used. While these can be helpful in some cases, it is important to understand how a document is to be used, for example, links would be no use for a hard copy manual. Where there is an external link, it is good practice to reference it too, as the link may not be valid indefinitely. Links are something else that a proof-reader needs to check for, as they do not always work for other people or on other devices. Some organisations encourage links; others discourage them. It is good to have a clear and consistent policy on this point.

Many of these formatting techniques can be incorporated in a template (see Section 2.8), so there is no need to address them on every document. A template should include a built-in automated table of contents and tables of figures and tables if appropriate. See Table A.1 for a template checklist. Putting time and effort upfront into a well-designed template is a good investment. It can save time and effort on every single use of the template and gives a consistent presentation.

Summary of points to include in a template

- Include a table of contents, unless fewer than ten pages.
- Consider tables of tables and figures.
- Adopt a consistent style on paragraph settings.
- Have a policy on using links: within a document, to external links or none.

5.3. Bullet points

I am often asked about the 'right' way to do bullet points (which are often abbreviated to 'bullets') or numbered lists. For example, should they start with capital or lower-case letters, and should they have punctuation after them or not? Most style guides largely agree on the details here. However, there are some differences in practice, and there are no universally agreed 'rules' on punctuating bullets. Style guides differ and some offer multiple options. As for other points of style, the best practice is to adopt a standard style and use it consistently.

Vertical lists in bullet point format are clearer and easier to read rather than separating a list of items by commas (or semicolons). Lists should be left aligned on the page. In the following example, the presentation in (b) is easier to read than (a) as it is in a vertical list.

(a) The Civil Engineering degree first year modules are mathematics, computational tools, highway engineering, soil mechanics, structures and environmental engineering.
(b) The Civil Engineering degree first year modules are:
    - mathematics
    - computational tools
    - highway engineering
    - soil mechanics
    - structures
    - environmental engineering

There should always be an introduction or lead-in line to bullet points. I often see a heading immediately followed by bullet points, which is not ideal. The purpose of bullets is to expand on something in a list format. Having no introduction leaves the reader guessing the purpose of the list and what they are supposed to do with it. A short sentence or phrase will suffice to introduce the bullets, and it may be appropriate to finish with a colon. Again, there is no universal agreement in style guides on the punctuation to introduce bullets. I have suggested ending the introduction with a colon as standard for engineering documents (see Table A.1), though full stops could be used, or even nothing.

Bullet points are usually indented from the introductory text with a symbol. These may be dashes, small circles, squares or other shapes (filled or hollow), and sub-bullets may be indented further, usually with a different symbol. Formatting of bullet point symbols and spacing can be a fiddly thing to check for consistency and is useful therefore to include in a template.

Bullet point lists should be similar in length. They should also be consistent with the same class of word, for example nouns (usually without an article) or verbs. If the latter, use the same form of the verb, for example the infinitive (to cover) or the imperative (cover).

Numbered lists may be used instead of bullet points, although this might suggest an order, priority or sequence in time. Use numbers where a sequence is required, for example, following a step-by-step procedure. An alphabetical order may be used for bullet points where there is no other specific order (see Section 3.5).

There is a broad consensus in style guides that bullet points should be introduced, that the lead-in should finish with a colon, and that they are indented from the introductory text. There is also general agreement that where bullet points are complete sentences, they should start with a capital letter and end with a full stop.

One debatable point is whether to use upper-case or lower-case letters to start bullet points. Most style guides recommend using lower-case letters for words and phrases that are not complete sentences. The exception here is for proper nouns, which should always start with a capital letter. However, the use of lower-case letters is not universally agreed.

The Oxford Manual of Style (Ritter, 2002) notes that fragments 'are usually (but not exclusively) lower-case', the rationale being that the bullet points are, in effect, continuing the sentence from the lead-in line. However, in practice, it is not unusual to see capital letters introducing bullet points of words and phrases, in addition to complete sentences. This may be a conscious choice, to be consistent with all bullet points.

One reason for capitalising all bullet points may be how Microsoft Word works. When writers want to add bullet points, they typically start a new line, then convert the text to a bullet point. However, Word default capitalises the first word on a new line, so if that is then made into a bullet point, it remains as a capital letter. It may be that writers may then accept this as standard practice. If a word is changed to lower case before it is converted to a bullet point, it will remain as lower case. Alternatively, the first bullet point may be changed to lower case. Either way, when further bullet points are added, they will be in lower case.

Another view is, 'If the sentence before the colon and the bullets after it form a continuous sentence, each line starts with a lowercase letter and there is no punctuation [...] If the bullets do not follow on in a continuous sentence, each line starts with an uppercase letter and there's no punctuation' (University of Nottingham, 2026).

A further point that style guides do not agree on is punctuation for bullet points and there are even multiple options within the same guides. OUP style (Ritter, 2002) is to punctuate displayed lists as if they were not separated. APA style (APA, 2020) has two options for bulleted lists for phrases: with or without end punctuation. Where there is punctuation, APA says to use a comma or a semicolon after each bullet point and a full stop after the final bullet.

A further style guide (Atherton, 2022) says bullet points for phrases may end with a semicolon, comma or nothing, and to always put a full stop at the end of the last bullet. Yet another (Cutts, 2013) always puts a full stop at the end of the final bullet. Cutts then gives multiple options for punctuating the other bullets: have no punctuation; for the penultimate bullet only, have a comma and add the word 'and' (', and'); or, for more legal documents, use a semicolon after all bullets, then the word 'and' ('; and') for the penultimate bullet.

Punctuation of bullet points is therefore one area to adopt a style and use it consistently. It may be helpful to include bullet points either in a style guide or in a template, showing examples in italics or hidden text (see Section 2.8). See also Table A.1 for a template checklist.

Author's note: My suggestion in Table A.1 (and the examples given in this Section 5.3) is to end bullet point introductions with a colon. I also recommend that bullet points that are words and phrases start with a lower-case letter and have no punctuation. Note that this book is printed in the Emerald publishing style, which has no punctuation at the end of the lead in line and a full stop after the final bullet point.

Bullet points summary:

- Always introduce bullets, never just heading then bullet point.
- End introduction or lead-in line with a colon (or follow style guide).
- Be consistent in bullet length and word class and form.
- For complete sentences, start with a capital letter and finish with a full stop.
- Start fragments with lower-case letters (or capital letter if on template or style guide).
- Do not add punctuation to bullet points (or follow style guide).

5.4. Numbers

Engineers communicate a lot of information in numbers, and this needs to be clear and consistent to avoid any ambiguity or misunderstanding. Numbers are clearer and easier to read in a vertical list or bullet points, rather than listed with commas. A table is best for a quantity of data. As for bullet points, there is general agreement about what is best practice, though not universal agreement. A couple of areas of disagreement are the decimal marker and the thousands separator. First, let us address the points where there is agreement.

The presentation of numbers is addressed in many style guides (APA, 2020; Office for National Statistics, 2026; Ritter, 2002). One key reference for numbers and units is the International System of Units (SI), which was established in 1960 (Bureau International des Poids et Mesures, 2025). There are also other standards and books that cover numbers in more detail (BSI, 2020; Gu, 2025; Hering, 2025).

When writing, do not start a sentence with a number in figures. Rephrase the sentence to avoid doing so or write the number in words. If writing in general terms, write the number in words, e.g. thousands. In SI, a billion is a thousand million (i.e. a billion is 10^9).

A common convention is to write out numbers below ten in words, e.g. zero, three, seven (APA, 2020; Office for National Statistics, 2026). Numbers above ten should then be written out in figures, e.g. 17. OUP style recommends this in technical contexts, but to use words for numbers below 100 in non-technical contexts (Ritter, 2002). Within a sentence, the same style in figures should be used consistently, e.g. '7 to 17'.

The decimal marker is the symbol used to show numbers, or parts of numbers, that are less than one (the decimal part). In the UK, the USA and most other English-speaking countries, common usage is to have a decimal point, i.e. a full stop, for example, 3.14. However, that usage is not universal and different formats can be confusing in an international business environment, for example in handling data files. SI allows either a point or a comma, 'The decimal marker chosen should be that which is customary in the language and context concerned' (Bureau International des Poids et Mesures, 2025).

Where the positive or negative signs are placed before a figure, there is no space between the sign and the figure, e.g. +3. There are no spaces between any type of brackets and characters in them (in words or figures), which is similar to quotation marks.

SI recommends, if a number is between –1 and +1, to use a zero before the decimal marker, e.g. –0.456 not –.456 (Bureau International des Poids et Mesures, 2025). This makes it clear, as a decimal point might be missed. However, although guides mostly agree on this point about adding a zero, there is not universal agreement. APA makes a distinction as to whether a statistic can exceed one or not; if it can exceed one, use a zero before the decimal point; if it cannot, such as a proportion, do not use a zero before the decimal point (APA, 2020).

Alongside the decimal marker, another point of difference is the thousands separator, which may be a comma or a space. These two issues are linked, as the comma must only be used for one or the other. It is easier to read large (or very small) numbers when the figures are separated in groups of three. SI defines prefixes that help present numbers without unnecessary digits (see Section 5.5). However, there may still be four or more digits.

SI uses spaces for groups of three (Bureau International des Poids et Mesures, 2025), and spaces may be used before and after the decimal marker. SI does not permit the use of dots or commas between groups of three. OUP style recommends a comma for non-technical work but a space for technical and scientific work (Ritter, 2002). Others (Gu, 2025) note that the SI system advocates a space, but that there are three ways to write the thousands separator.

The comma is used in most English-speaking countries, and the period (full stop) in many non-English-speaking countries.

Note that technical drawings, even in the UK, may not follow the same convention. Engineers may be aware of the standard BS 8888 (BSI, 2020), which is widely used for geometric dimensioning and tolerancing (GD&T). The number format in that standard is to use a comma for the decimal marker, and a space for the thousands separator. For example: 16,5 means sixteen and a half; 12 000 means twelve thousand. Two decimal places would be shown after a comma, e.g. 17,03. Although possible, it is rare to see a dimension or a tolerance to more than two decimal places on a technical drawing. It is unlikely therefore that someone would mistake this decimal marker comma for a thousands separator, even if they were not familiar with BS 8888.

For years, the thousands separators is not used, e.g. 1987 and 2026. For decades, numbers are preferred to words and do not have apostrophes (the single closing quote), e.g. 1920s, unless they are abbreviated, e.g. '20s (Ritter, 2002). Where words are used, they should be lower case when referring to millennia, centuries and decades, e.g. 'nineteen-twenties' with a hyphen. Write out dates in full to avoid confusion between UK and US date format. For example, 4 July 2026 would be 4/7/26 in the UK, but 7/4/26 in the US.

For telephone numbers in the UK reader, break up the number with a space, e.g. 01925 123456. For an international reader, add the international dialling code and remove the leading zero, e.g. + 44 1925 123456. As for the thousands separator, it is much easier to read numbers with some spaces.

When using the percentage symbol, the SI manual (Bureau International des Poids et Mesures, 2025) recommends using a space between the number and the symbol, such as 50 % and 80 %. However, most other style guides have no space, so would write 50% and 80% (APA, 2020; Gu, 2025; Ritter, 2002). I suggest using no space with the percentage symbol, as that is the most common usage. Consider the guidelines for percentages as similar to those for currencies.

By contrast, for units (see Section 5.5), I would recommend using a space between the number and the unit. There is much more of a consensus on using that space, and it reads more clearly too. Even so, I am aware that some company styles have no spaces before units. As for most elements of style, choose a style and use it consistently.

In accounting, a comma is used for the thousands separator and the decimal point for the decimal marker. For currencies, there is usually no space between a currency symbol and the amount in figures, e.g. £1000 and $1000. Nor should there be a space between a figure and pence (or cents), e.g. 50p. Be consistent within a sentence, for example, £1000.00 and £0.50.

Be careful not to present figures that might imply more significant figures than the accuracy of the available date. Engineers often work with many variables and use factors of safety. Pi (π) is 3.14 to two decimal places (three significant figures). If greater accuracy is required, pi is 3.142 to three decimal places (four significant figures).

Numbers summary

- Follow standard conventions for writing numbers in words or figures.
- Between −1 and +1, use a zero before the decimal marker (or follow a style guide).
- Use a decimal point for the decimal marker (or follow a style guide).
- Use spaces for the thousands separator (or use commas).
- For technical drawings: use a comma for decimal marker and a space for thousands separator.

5.5. Units

As for numbers, units should be used clearly and consistently. There are guides, standards and books for units as there are for numbers. There is the International System of Units (SI) brochure (Bureau International des Poids et Mesures, 2025), which is available free online. There are also other references (Gu, 2025; Hering, 2025).

Use the standard SI names or symbols consistently and do not mix names and symbols, such as metre and m. Do not use any other abbreviations in either words or units, e.g. do not use ‘sec’ for second, or the word ‘per’. When using imperial units, for example with US clients or legacy equipment, follow standard symbols and conventions.

Always use a space between the number and the unit, e.g. 10 mm. This is recommended by SI (Bureau International des Poids et Mesures, 2025) and others. This includes a space between a temperature and the unit ‘degree Celsius’, e.g. 15 °C (note there is a space between the figure and the degree symbol, but not between the degree symbol and the letter C). By contrast, for an angle, there is no space between the figure and the degree symbol, such as 90°. There should also be a space before and after mathematical operators, such as +.

Do not write units in plural. Although engineers might say out loud ‘ten kilograms’ with an ‘s’, it should only ever be written as 10 kg. In SI, the letter ‘s’ denotes a second, which is the unit of time.

Unit symbols are lower case unless derived from a proper name, so s for second and m for metre, but A for ampere and K for kelvin. Always use the standard SI units where possible. Do not mix words and units in the same expression. Units are not abbreviations, so are not followed by a full stop unless they are at the end of a sentence.

SI defines prefix names and prefix symbols for multiples or sub-multiples of units, e.g. gigahertz and nanometre, or GHz and nm. In words, all these prefixes are lower case, unless they are at the start of a sentence. As they are prefixes, there is no space between the prefix and the rest of the word.

These are in multiples of three, e.g. kilo-, mega- and giga-. Using the right multiple means that most numbers can be presented in the range of 0 to 1000, e.g. 500 GW, which is clearer and easier to read than 500 000 MW or 500 000 000 W.

Take care with upper and lower case for prefixes, mega (M) is 10^6 and milli (m) is 10^{-3}. One common error is in kilo, which is lower case, so the unit of power is kW (in SI, a capital letter K is used for kelvin, the unit of temperature).

When writing a tolerance, the units should be written for all figures, e.g. 100 m ± 10 m, unless both figures are contained within the same brackets, e.g. (100 ± 10) m.

5.6. Equations

SI recommends using a multiplication sign (×) for numbers, but allows various forms for symbols, e.g. ab, $a\ b$, $a{\cdot}b$ or $a \times b$. Similarly, for division, SI recommends a solidus, which is a technical name for a slash (/) for values, and brackets to avoid any ambiguity. Symbols may be written in several different ways: a/b, a/b with a horizontal line or $a\ b^{-1}$. As for any other style choice, adopt a style and use it consistently, e.g. use a space in multiplying symbols or do not.

Equations should be written in an editable form (not a picture). Microsoft Word does have an equation function where they can be typed in. For more complex equations, there are equation editor programs, like MathType or Overleaf, an online LaTeX editor. As for other aspects of writing, choose the technology tools that best suit the purpose (see Section 2.6), and use those consistently across the organisation.

However, these are editing tools for presenting equations not doing the maths. There is other software available, such as MATLAB or Mathcad, that will do programming or solve mathematical problems. If appropriate, reference from Word to where the technical work is done.

Ensure that a technical checker can follow equations so that they are able to check them. For example, a screenshot of calculations done in Excel that refer to other cells cannot be checked without the original Excel workbook.

For proof-reading, ensure that all characters are legible, including all the symbols and superscripts. As for figures and tables, it helps the reader if equations are numbered and referred to in the body of the text. Microsoft Word can caption equations in a similar way to tables and figures.

As described in Section 5.5, always define units and do not leave it to the reader to assume or guess them. Use of SI units is best practice in the UK and many other countries. It also does more. The seven base quantities of the SI system (Table 5.3) cover the seven dimensions. These help with understanding mathematical problems and ensuring that they are solvable. Dimensions cannot be converted, for example a mass and a length cannot be added together. Units can be converted, providing that they are the same dimension, for example lengths in millimetres and metres.

Table 5.3 SI base units

Base quantity	Name	Symbol
Length	metre	m
Mass	kilogram	kg
Time	second	s
Thermodynamic temperature	kelvin	K
Amount of substance	mole	mol
Electric current	ampere	A
Luminous intensity	candela	cd

All engineering properties are derived from the seven base SI units. Like a family tree, all the other properties and parameters in science and engineering are made by combining these in different ways, which give combined units, for example:

$$\text{velocity}(V) = \text{displacement }(x) / \text{ time }(t) = \text{m/s}$$

This can be an important check as equations become more complex. Although some of these parameters are given special names, such as the newton (N) as the unit of force, they can still be derived from (or broken down to) their base SI units:

$$\text{force}(F) = \text{mass}(m) \times \text{acceleration}(a) = \text{kg m/s}^2 = \text{N}$$

So, making sure that all figures have units not only makes the writing clear and consistent but also ensures that the equation is valid, i.e. complete and correct.

5.7. Tables

Tables are useful to present large amounts of information, often numerical data. Tables should have a specific purpose and be presented clearly so that they could stand alone. Most of the content in this section on tables also applies to figures (see Section 5.8), so for brevity the word ‘figures’ is not repeated. Oxford and APA styles provide further guidance on presenting figures and tables (APA, 2020; Ritter, 2002).

Consider what level of detail is necessary for the purpose of a table. What data and information are needed for the main ‘story’, and what is supporting material that could go in an appendix (or be omitted)? For example, the results section of a test report might have one summary table, and twenty individual test outputs could go in an appendix.

In a formal document, tables should be labelled. An exception is when an appendix contains only one table; then the appendix title is sufficient without a further label (APA, 2020). It is also good practice to bring the reader’s attention to the table in the body of the text. This makes sure that it is adding value and helps the reader know when to look at it. Captions should be brief, within one line. A list of tables may also be useful (see Section 5.2).

Signposting of tables is a point of detail that helps the reader navigate a document. It also helps to explain the purpose of the table. I have seen many examples of documents that are mostly in tabular form, sometimes with section headings followed by tables with no text to explain them. This can be quite off-putting to a reader. Just a few words will help orient the reader and guide them, so there is no assumed knowledge or need to work out what to do with the table. For example, in a specification, an introductory sentence could explain that a table contains requirements for a particular sub-system.

Often tables are embedded within a document, although some organisations prefer them all together at the end. If a table is embedded in the main document, it should be close to the text that refers to it. Oxford style (Ritter, 2002) says preferably following that text. Refer back to a previous table rather than repeating data. This avoids the risk of inconsistency (and is more concise). Using references such as, 'Table 1 shows' rather than, 'the table below shows' helps avoid potential problems when there is a page break in the final version of the document, and the table is on the next page. Writing the table number also helps people using screen readers.

A common convention, and also Microsoft Word's default, is to put labels above a table or equation and below a figure. This is a style choice, and some prefer to put all labels in the same position. This is a point of detail where an example can be given in a template or a style guide. Microsoft Word calls these labels 'captions', and they can be used for figures, tables and equations. They are added from the Reference tab, under 'Insert Caption'.

The default in Microsoft Word is to label all tables consecutively, for example 'Table 1', 'Table 2' and so on. There is also an option in Microsoft Word caption numbering to add the chapter number. For example, Section 5 of a document may have 'Table 5.1', 'Table 5.2' and so on. This may be helpful for a large document. However, as for any point of style, consistency is key; if captions are defined in an organisation's template, follow that.

Tables should be presented in a consistent format, not only in the captioning but also in the content. All columns should be labelled, and units defined. Numbers in a column should be in line and presented to the same number of decimal points. Keep a table on one page where possible; if a table spans more than one page, repeat the headings. Define abbreviations in tables even if they are explained elsewhere, except for those commonly used.

5.8. Figures

Figures may cover a wide range of visuals, including photographs, charts and drawings. Some terms are used loosely and interchangeably, like charts and graphs. I am treating a graph as a specific type of chart with numerical data, for example, showing relationships or plotted with axes. Often, engineers are graphically representing data in some form of graph or chart.

As for other aspects of writing, the term of the specific type of figure is not important. What does matter is the purpose of that figure for the reader. For example, a detailed wiring diagram might be to enable an electrician to install cables; an outline sketch drawing in a user manual might be for a non-technical person to connect up a consumer product.

A key principle is that one figure should make one point. For multiple points, use multiple figures. Fewer words are required when using figures, so these should be planned in alongside paragraph bullet points when starting to write (see Chapter 3). Figures should not be too busy or complex, as this makes them harder to understand and may distract the reader from the main point.

Another key point is to choose the best type of chart for the purpose. To understand this, a good place to start is the type of information. One way to categorise engineering information is as guidance only, symbolic or schematic, and prescriptive (Matthews, 2012).

> Do I need to give the answers, to check my answers, or to look for answers?
>
> (Scott Berinato)

Another way to classify information follows from two questions (Berinato, 2016), which are: is it 'conceptual' or 'data-driven', and is it 'declarative' or 'exploratory'? Conceptual is when there are no data, contrasted with data-driven where data are available. Declarative is when something is be presented, contrasted with exploratory, when something is to be found out. The combination of these two questions gives four types of visualisation.

1. Conceptual exploratory may be hand sketches in workshops to generate ideas.
2. Conceptual declarative may be flowcharts or organisation charts that tell people something.
3. Data-driven declarative is often presenting technical information.
4. Data-driven exploratory is finding out detailed information. There are two levels of data-driven exploratory information.
    a. Confirmation might be testing a hypothesis; for example, is there a relationship between two parameters?
    b. Exploration might be more open-ended, often with statistical techniques.

This approach helps to clarify not only the purpose of the chart, but also to select the best type of chart and software tool for that figure. Engineers may use many different types of charts and software packages, such as flowcharts, *x*–*y* scatter plots or statistical analyses.

Most of the good-practice guidance on tables (see Section 5.7) applies to figures, too. For example, figures should be labelled and referenced in the text. All points of detail should be labelled, for example *x*-axis, *y*-axis, scale and units, and a legend for symbols or lines. 'Callouts' may be used to highlight a particular issue, for example, drawing a balloon around a point of interest.

All figures should be clear and legible, with sufficient resolution and contrast, so that they are accessible to all (see Section 5.9). A practical tip is to avoid relying solely on colour to distinguish between lines or segments. Line types can be changed from solid to dotted, segments can be filled or hatched. This may help people with visual impairments, such as colour blindness. Also, if documents are viewed or printed in black and white, different colours may appear in the same shade of grey.

There are ethical issues to be aware of with figures. The scale used must not be misleading, implying a relationship that is not valid. It should always be clear if a picture is an artist's impression or AI-generated, rather than an original photograph. When using figures containing other's IP, ensure that permission is given by the copyright holder (see Section 3.8).

5.9. Diversity and inclusion

> All members shall contribute positively to the culture and public perception of the profession and the Institution: [...] They shall support and encourage equality, diversity and inclusion.
>
> (IMechE, 2021)

This section focuses on how to be inclusive in technical writing to better help the reader. It also touches on not only writing for our reader but also how we write about people.

Firstly, we need to be sensitive about how we write about people to be inclusive. For example, avoid collective terms such as 'the disabled' or 'the blind', and instead use 'disabled people' or 'people with visual impairments' (Disability Unit, 2025). Guidelines in different parts of the world may vary. In the UK, 'disabled people' is often preferred over 'people with disabilities', which is used more globally (Business Disability Forum, 2025).

> Making information easier to understand for people with learning difficulties is about more than making your text size bigger and putting some symbols or pictures in your document.
>
> It's about working with people your information is for, finding out together how you can make the information useful and accessible for them. It's about thinking about how we all make sense of information.
>
> EasyInfo (quoted in Department of Health, 2010)

The National Autistic Society recommends asking the person or group how they would like to be described, and otherwise recommend saying 'autistic person' rather than a 'person with autism'. (National Autistic Society, 2025). Avoid using the word 'disorder' unless specifically referring to the medical terminology of 'autism spectrum disorder'. For dyslexia too, avoid using the word 'disorder' or 'disability' (Davis Dyslexia Association International, 2025). The Davis Dyslexia Association International also recommend saying 'dyslexic person' rather than 'a person with dyslexia'.

This sensitivity applies in other contexts too. For example, in the Caribbean, they prefer to say African American people 'were enslaved' or were 'enslaved people', rather than 'slaves'. That identity does not define the totality of a person. If unsure of the right language to use, ask. Use neutral language, like 'condition' rather than 'issue' or 'problem'.

The British Dyslexia Association have a dyslexia style guide (British Dyslexia Association, 2014), which covers readable fonts, headings and structure, colour, layout and writing style. The good news is that many of these elements of style also help everyone else to read more easily too, such as using the active voice and simple clear language. Some of the guidelines on font and structure can be incorporated in a template, for example, 'For headings, use a font size that is at least 20% larger than the normal text' and 'left align text, without justification'.

Other guidelines for dyslexia say to avoid hyphens, which are more commonly used with justified text. Whilst published documents are usually justified and use hyphens, for internal engineering documents left aligned text without hyphens may be easier to read.

The Check Accessibility function in Microsoft Word Review checks for many issues, such as insufficient contrast or missing 'alt text', which is alternative text for a figure. Ensuring that a template is accessible reduces the amount of revisions required for documents (see Section 2.8).

There are also the Web Content Accessibility Guidelines (WCAG) 2.2 (W3C, 2024), which largely cover other aspects of web accessibility. However, from a writing perspective, there is a section on 'understandable and readable', including the reading level (see Section 5.1). It says to 'make text content readable and understandable' (W3C, 2024).

5.10. House and other style guides

There are published style guides, such as *The Oxford Manual of Style* (Ritter, 2002), which is sometimes referred to as OUP style, and the APA *Publication Manual* (APA, 2020). I understand that there are also the *Chicago Manual of Style* (University of Chicago Press, 2024) and the *MLA Handbook* on style (MLA, 2021), although I am not personally familiar with either of these. There are many other books on style, such as *The Elements of Style* (Strunk and White, 2000) and *The Economist Style Guide* (Grimond, 2013).

There are also online style guides. The UK government and its associated bodies have some online style guides (Government Digital Service, 2026; Office for National Statistics, 2026). There are some Ministry of Defence guides, although most of these require log-in access. Many publishers and publications, such as newspapers, will also have their own style guides.

While there are plenty of style guides and common conventions on many elements of style, there is no universal agreement. For example, as we have seen, there is no agreed rule as to how to punctuate bullet points (see Section 5.3). Another issue is how numbers are written (see Section 5.4), which is more than presentation, as it could lead to misunderstanding.

Engineers often use engineering standards that may contain options. As such, engineers need to be familiar with any standard they reference, as they may need to select options (or note specific conditions). Also, where there are multiple standards, they should be ranked, in case there is any inconsistency. Style guides are similar, in that they may contain acceptable options, and there may be inconsistencies between them.

Many of these style guides cover details only of interest to publishers, such as editorial mark-up, spacing and typography. Engineers should not need to be familiar with that level of detail for their everyday work. However, published guides may be a useful reference source for an organisation in preparing an in-house style guide, or to supplement one where an element of style is not otherwise addressed.

The appendix provides a basis for a style guide, broken down into three tables for ease of use. Many aspects of style can be incorporated in a template, either in the main content or as example text in italics or hidden text. Table A.1 gives an expanded checklist for a template, including formatting issues, such as bullet points. Where such issues are not addressed in a template, they can be added to a style guide checklist. Table A.2 is a 'formal and clear' checklist, covering elements of style that are commonly accepted, such as no slang. Table A.3 supplements this, highlighting issues where an organisation needs to make a style choice, such as captions above or below a table.

I would recommend all organisations adopt a house style guide. It could be based around one of the published style guides, with preferred options. Alternatively, it could be developed from the checklists in the appendix, creating a template and an agreed style guide checklist. An in-house style guide can help an organisation to define a clear and consistent, professional style and communicate it to everyone. It also saves time for the writer and the reviewer, avoiding edits for style issues and debates on the 'right way' to do something.

REFERENCES

APA (American Psychological Association) (2020) *Publication Manual of the American Psychological Association: The Official Guide to APA Style*, 7th edn. APA, Washington, DC, USA.

ASD (Aerospace, Security and Defence Industries Association of Europe) (2025) *ASD-STE-100. Simplified Technical English: Standard for Technical Documentation*, 9th edn. ASD, Brussels, Belgium.

Atherton T (2022) *Technical Report Writing and Style Guide: How to Write Even Better Technical Reports*, 7th edn. Amazon, Seattle, WA, USA.

Berinato S (2016) *Good Charts: The HBR Guide to Making Smarter, More Persuasive Data Visualizations*. Harvard Business Review Press, Boston, MA, USA.

British Dyslexia Association (2014) *BDA Dyslexia Style Guide*. British Dyslexia Association, Bracknell, UK. https://www.bumc.bu.edu/jmedday/files/2025/05/Dyslexia-Style-Guide-2023-BDA-Style-Guide-2023.pdf (accessed 06/05/2026).

BSI (2020) BS 8888:2020: Technical product documentation and specification. BSI, London, UK.

Bureau International des Poids et Mesures (2025) *The International System of Units (SI)*. https://www.bipm.org/en/publications/si-brochure (accessed 13/01/2026).

Business Disability Forum (2025) Inclusive Language: What It Is and Why It Is Important. https://businessdisabilityforum.org.uk/resource/inclusive-language-what-it-is-and-why-it-is-important-2/ (accessed 07/05/2026).

Cutts M (2013) *Oxford Guide to Plain English*, 4th edn. Oxford University Press, Oxford, UK.

Davis Dyslexia Association International (2025) Describing dyslexia. *Dyslexia the Gift Blog*. https://blog.dyslexia.com/about-us/guest-contributors/guideline-for-posting/describing-dyslexia/ (accessed 07/05/2026).

Department of Health (2010) Making Written Information Easier to Understand for People with Learning Disabilities: Guidance for People Who Commission or Produce Easy Read Information. Department of Health, London, UK. https://assets.publishing.service.gov.uk/media/5a74adb4e5274a56317a6411/dh_121927.pdf (accessed 12/12/2025).

Disability Unit (2025) Inclusive Language: Words to Use and Avoid When Writing about Disability. Cabinet Office, London, UK. https://www.gov.uk/government/publications/inclusive-communication/inclusive-language-words-to-use-and-avoid-when-writing-about-disability (accessed 07/05/2026).

Government Digital Service (2026) Style Guide. https://www.gov.uk/guidance/style-guide (accessed 28/02/2026).

Grimond J (ed.) (2013) *The Economist Style Guide: The Bestselling Guide to English Usage*, 10th edn. The Economist, London, UK.

Gu Y (2025) *Engineering Dimensions, Units, and Conversions*. CRC, Boca Raton, FL, USA.

Gunning R (1968) *The Technique of Clear Writing*, revised edn. McGraw-Hill, New York, NY, USA.

Hering H (2025) *How to Write Technical and Scientific Reports: Understanding Structure, Good Design, Convincing Presentation*, 3rd edn. Springer, Berlin, Germany.

IMechE (Institution of Mechanical Engineers) (2021) *Code of Conduct Regulations*. IMechE, London, UK. https://www.imeche.org/about-us/imeche-governance/governance-and-finance-reviews/code-of-conduct-explained (accessed 13/01/2026).

Matthews C (2012) *Engineers' Data Book*, 4th edn. Wiley, Chichester, UK.

MLA (Modern Language Association of America) (2021) *MLA Handbook*, 9th edn. MLA, New York, NY, USA.

National Autistic Society (2025) *How to Talk and Write About Autism*. National Autistic Society, London, UK. https://www.autism.org.uk/contact-us/media-enquiries/how-to-talk-and-write-about-autism (accessed 07/05/2026).

Office for National Statistics (2026) *Content Style Guide*. https://service-manual.ons.gov.uk/content (accessed 07/05/2026).

Ritter RM (ed.) (2002) *The Oxford Manual of Style: The Essential Handbook for All Writers, Editors, and Publishers*. BCA, London, UK.

Seely J (2013) *Oxford Guide to Effective Writing & Speaking: How to Communicate Clearly*, 3rd edn. Oxford University Press, Oxford, UK.

Strunk W and White EB (2000) *The Elements of Style*, 4th edn. Pearson, Upper Saddle River, NJ, USA.

University of Chicago Press (2024) The Chicago Manual of Style, 18th edn. University of Chicago Press, Chicago, IL, USA.

University of Nottingham (2026) Bullet Points. https://www.nottingham.ac.uk/brand/styleguide/?id=6869484f-86d1-4fff-9b7c-f02c511a84d0 (accessed 04/03/2026).

Van Emden J and Becker L (2018) *Writing for Engineers*, 4th edn. Bloomsbury, London, UK.

W3C (World Wide Web Consortium) (2024) Web Content Accessibility Guidelines (WCAG) 2.2. https://www.w3.org/TR/WCAG22/ (accessed 04/03/2026).

Technical Engineering Writing: A Practical Guide for Professional Engineers

emerald PUBLISHING ice Publishing

Jean Billingsley
ISBN 978-1-83662-927-6
https://doi.org/10.1108/978-1-83662-924-520261007

Chapter 6
Writing common engineering documents

The same basic principles apply to writing any engineering document, particularly around planning and focusing on a specific purpose and reader. As covered in outlining PRIDE, it is also important to integrate a document as part of the work and then, when writing, first draft then edit. This approach applies for engineers writing anything. This chapter covers some specific points relating to a selection of common types of engineering document.

6.1. Technical reports

Reports of one form or another are probably the type of document engineers write most often and the majority of the standard guidance applies here. There are a variety of different types of report, and some describe reports as all-encompassing, including other types of documents such as proposals.

Typical contents for a report are

- summary or abstract
- introduction
- method
- results
- discussion
- conclusion.

A report is intended to be read from start to finish so it should have a start, middle and end. This might be described as 'telling a story', and the structure should suit its purpose. Often the structure is chronological, but it may follow another way to organise ideas (see Section 3.5).

The summary is placed at the start, although it should be written last, as it summarises the completed document. Reports should always have a summary, even for a short report, as it is a really important section. The writer is the best person to summarise a report and communicate the key message.

An introduction should contain the problem statement, and enough background for the target reader. A method describes what was done. This needs to be sufficient that a competent peer could reproduce the work, including: source of input data, equipment and tools used and any assumptions made. Results should be factual and evidenced. A discussion, if necessary, may interpret the results within the writer's field of expertise. Conclusions should logically follow from the previous sections. There may or may not be recommendations for action. There are usually references and may be an appendix.

6.2. Specifications and instructions

There is a lot more work to preparing a specification than the writing, including stakeholder engagement, using standards and the project management stages. This section is focusing on the writing aspects only. The structure of specifications and instructions varies, depending on the type of specification (see Section 3.5). Unlike reports, specifications do not usually have summaries, and so the introduction is an important section. It should state the objective and give a brief summary of the technical contents.

Writing specifications or other types of written instructions such as procedures or technical manuals needs to follow all the good practice for technical engineering writing covered so far. However, it needs an even tighter, shorter style of writing. For requirements engineering, it needs to be unambiguous and there are some prescriptive standards available.

The same words should be used for consistency, rather than using a variety of words as for other engineering documents. In a specification it would not be good practice to refer to the same person or thing in multiple ways in different sections, such as: 'the contractor shall…', 'the tenderer shall…' and 'the supplier shall…'. The reader might be confused as to which requirements applied to them. For a product specification, the word 'shall' indicates a requirement. This is a widely accepted standard.

For procedures or instructions, start with the verb in the imperative (mandative case).

1. Isolate the system.
2. Remove the cover.
3. Discard and replace the seal.
4. Replace the cover.

Use only one clause in these sentences, unless there is a condition.

> When the light comes on, press the green button.

In specifications, do not use brand or company names. In everyday life, we use words like hoover, sellotape and post-it. However, Hoover, Sellotape and Post-it are company or brand names. Similarly in engineering, 'tannoy' is used generically to mean any public address system. However, Tannoy is a company that makes loudspeakers and public address systems, and Tannoy is a registered trademark. Swagelok is trademarked, and written as Swagelok® (with a small capital letter R in a circle) and the company supplies fittings, valves and other components and systems. Portakabin and Tarmac are also registered trademarks.

6.3. Proposals and business cases

Proposals and business cases are similar: typically a proposal is for an external client, and a business case is an internal justification to make an investment.

As for writing any engineering document, proposals should be written with the reader in mind. This means understanding the prospective client and their problem, then tailoring the content to explain how the company can address their problem. A proposal listing products and services without explaining the benefit to the client may miss the mark. Proposals need a strong title, particularly as they need to persuade the reader. Contents may also include previous examples of similar experience to demonstrate a track record and external recognitions, awards or social proof, such as testimonials.

A proposal must be focused on the prospective client, and consider their questions: 'What's in it for me?' (WIIFM). Writing should use persuasive language, with useful words to know and use in bids (Manser, 2014) (see Section 4.2).

For a business case, the content is often financial. To justify R&D spend or a capital investment, forecast returns or savings are required. If a project spans multiple years, a net present value treatment of income and expenditure is needed for a valid comparison to be made. There is usually also a strategic rationale, for example increasing revenue or maintaining sales with a change in legislation.

6.4. Journal articles

Research papers are often published as journal articles (or sometimes may be conference proceedings). There are many things to consider when deciding where to publish research, beyond the target readership. There may be timing issues, as conferences will often issue a 'call for proposals', and prospective authors need to submit a synopsis for consideration according to a fixed timetable. The reach or circulation of journals can vary greatly too.

Some journals may have a more rigorous academic approach, while others approach; others might be perceived as an easier route to getting published. Typically, journals will have editorial teams and any articles submitted will be reviewed by at least two academic peers. However, that is not always the case.

In addition to the author and the journal, another key stakeholder is the author's employer. It is often a measure of success for academic individuals and institutions to be published (and subsequently to be cited). Certain subjects are considered more academic than others, and some journals more prestigious than others. I recall when I was doing my MBA that project management was not perceived as a very academic subject and publishing in associated in relevant journals did not 'rank' as highly with universities.

While publication might be encouraged for publicly funded research, at other times there may be restrictions on publication. Not all funders will permit publication, or there might be very tight controls over what may be published; for example, there may be a private industry partner who is looking for a competitive edge. If a company is considering applying for a patent, publishing would be prohibited as that would mean the work had been disclosed into the public domain and a patent could not be granted.

In terms of readability, journal articles will typically be harder to read, as measured by the grade level in readability statistics (see Section 5.1). Compared with a target level of 9–11 for most engineering documents, such as reports, a grade level of around 14 would be a good target for an academic journal. This is because the writing is likely to include a higher average length of word, which is appropriate for the purpose, reader and context. The readers of journals are typically experts in a technical field, who are very familiar with the specialist terminology.

The format for submission is likely to be prescriptive in terms of word count and presentation. Journals often have a two-column format. APA has a set of standards for journal articles, the Journal Article Reporting Standards (JARS). The writing style in academic journals is more about effective argument and critical thinking, compared to reports (Swatridge, 2014). An extensive list of references is expected.

A paper needs a good title and an abstract, which is another name for a summary (the term usually used in industry). A paper may or may not have headings, but it still needs structure to organise the ideas. One of the models for research writing is of introduction, method, results, discussion and conclusion.

6.5. Writing for publication

For writing books, there is a process with a publisher. Similar to journals, different publishers will have their own specialisms. For engineers, there are some publishers who focus more on technical and educational books, such as Emerald/ICE Publishing. Publishers will have their own guidelines for authors, including style.

The author develops a book proposal together with a commissioning editor. This covers the topic, contents, target audience and other details. As covered under planning (Section 2.10), there may be extra roles not found in other engineering documents, such as editor, publisher, typesetter and professional indexer.

With Amazon Kindle/Create Space, self-publishing is straightforward. While this does not have the same prestige as a publisher, it can be an entry route to publishing for some individuals.

When writing for publication, there are usually quite prescriptive designs, and extensive style and publishing guidelines. Most published books and papers justify text (aligning text left and right to make all lines the same length) and therefore use hyphens to enable this. Internal documents such as reports do not need to do this, and they can align left only and not allow hyphens. Books usually have widow control too.

Many US English spelling variants are commonly accepted in UK English (see Section 4.4). Some publishers prefer a strict interpretation of UK English and will only use ‘organisation’ not ‘organization’, ‘analyse’ not ‘analyze’, and ‘learnt’ not ‘learned’. Other guidelines may include references for each chapter rather than the whole book.

6.6. Non-technical writing

Engineers may also write a range of other non-technical documents, which need to focus on the purpose and the reader.

Minutes of a meeting should be brief and not a verbatim transcript, and it is not appropriate to identify who said what. Minutes should be issued promptly after a meeting, note key points of progress and decisions made, and allocate actions and action owners, thereby giving people permission. Overall, the purpose of minutes is to record decisions and to progress work.

Progress reporting should follow the PRINCE2 project management principle of 'manage by exception'. It is not good use of time to write an extensive report that says that the project is progressing to plan. Readers appreciate brief highlights. For example, use a weekly 'four-block' dashboard of: achievements, risks, next steps and issues.

Some writing is focused on career and personal development, for example, a curriculum vitae (CV), an application for Chartered Engineer status or a job application. The purpose of these documents is to promote the engineer, so the content needs to be about their achievement and personal contribution, not a description of the projects worked on. For a CV, writing style is brief, maximum two pages, and uses strong verbs, such as 'designed' and 'led a team'.

Press releases sent to journalists use the style, structure and content that they are looking for, so that they are more likely to get published and with few changes. Press releases have a structure within one heading, similar to the BLUF or inverted pyramid (see Section 3.5). There is an attention-grabbing headline and start, then further details later for those who wish to continue reading. There is typically a direct speech quote.

An example structure for a press release

- company name
- date
- heading
- sub-heading
- lead paragraph (e.g. 5Ws)
- main story
- comments by people involved
- optional images, further interesting details, supporting information, background to the story, about the company
- contact details.

Some elements of style may be less formal for certain types of writing, so some style choices depend on the nature of the document (see Table A.3). For example, a direct speech quote is used in a press release or magazine article. Quotations in other forms of engineering writing are usually extracts from written documents.

> HMS Prince of Wales is one of the most powerful surface warships ever constructed in the UK. Her flight deck is 70 metres wide and 280 metres long – enough space for three football pitches – and she holds 45 days' worth of food in stores.
>
> (Royal Navy, 2025)

Another example is the use of general comparisons, like 'the width of a human hair'. These are not usually used in formal documents, as they are imprecise. However, they are a useful device to convey a sense of scale, ideally with the actual figures. Comparisons with double decker buses and football pitches are commonly found in newspapers. General comparisons may also be used in some forms of writing, such as grey literature.

Grey literature is a term relating to something that is not produced by commercial publishers. Unlike books and journals, these are not edited or peer-reviewed. This can cover a wide range of documents, such as policy documents, white papers and blogs. Conference proceedings may also be classed as grey literature, as they may not be peer-reviewed in the same way as journal articles.

'Thought pieces' may also come under this category of grey literature. For example, IMechE produce reports and policy statements (IMechE, 2026). These address engineering issues, aligned to their vision and themes, and make policy recommendations.

REFERENCES

IMechE (Institution of Mechanical Engineers) (2026) *En Route to Inclusive Public Transport: Roundtable Summary*. IMechE, London, UK. https://www.imeche.org/docs/default-source/1-oscar/reports-policy-statements-and-documents/roundtable-summary---en-route-to-inclusive-public-transport.pdf?sfvrsn=582c5811_2 (accessed 13/03/2026).

Manser MH (2014) *1001 Words You Need to Know and Use: An A–Z of Effective Vocabulary*. Oxford University Press, Oxford, UK.

Royal Navy (2025) HMS *Prince of Wales* (R09). https://www.royalnavy.mod.uk/organisation/units-and-squadrons/aircraft-carriers/hms-prince-of-wales (accessed 07/05/2026).

Swatridge C (2014) *The Oxford Guide to Effective Argument & Critical Thinking*. Oxford University Press, Oxford, UK.

Jean Billingsley
ISBN 978-1-83662-927-6
https://doi.org/10.1108/978-1-83662-924-520261008

Appendix
Style guide and checklists

Table A.1 Template expanded example checklist (continued on next page)

Element of style	Example(s)
Proofing language	UK English
Headings and sub-headings	Numbered in themes/styles
Title and organisation information	Included
Automatic table of contents	Always include (or only if 10 + pages)
Table of figures	Included (or not, or only if 3 + figures)
Table of tables	Included (or not, or only if 3 + tables)
Figures and tables	In body of report (or all at the end)
Font and size – header level 1	Aptos (display) 18 bold
Font and size – header level 2	Aptos (display) 14 bold
Font and size – header level 2	Aptos (display) 11 bold
Font and size – body text	Aptos (body) 11
References	Included (with example)
Appendix	Included
Glossary of acronyms & abbreviations	Included
Header (or footer) with document number and revision	Included
Footer with page number, copyright and confidentiality notice	Included
Tables captioned	Above table
Figures captioned	Below figure
Equations labelled	Above equation
First sentence of new section indented	Yes (or no)
Start of subsequent paragraphs indented	Yes (or no)
Line space between paragraphs	No (or yes if no in line above)
Paragraph widow and orphan control	On
Paragraph hyphen	Do not hyphenate

(*Continued*)

Table A.1 Continued

Element of style	Example(s)
Text justification	Left only (or left and right)
Sample text identified (if used)	Italics (or hidden or *lorem ipsum*)
Bullet point examples	Italics (or hidden or *lorem ipsum*)
Bullet point introduction	Ends with a colon (or full stop or nothing)
Bullet point complete sentences, start	Capital letter
Bullet point complete sentences, end	Full stop
Bullet point words or phrases, start	Lower-case letter (or capital letter)
Bullet point words or phrases	No punctuation (or commas or semicolon)
Bullet point words or phrases, penultimate	No punctuation (or commas, semicolon, comma and, or semicolon and)
Bullet point words or phrases, last one only	No punctuation (or full stop – always full stop if other bullet points are punctuated)

Bullet and sub-bullet point symbols (also incorporating four lines above)	Example fragment bullet points:
	■ first bullet point
	■ second bullet point
	■ third bullet point
	   o sub-bullet point one
	   o sub-bullet point two

Links permitted	None (within a document or external)
Template checked for accessibility	Contrast, alt text

Table A.2 Formal and clear checklist

Element of style to avoid	Example(s)
First person	I, we
Personal pronouns	She, they
Contractions	Can't
Ampersand	&
Feeling	I feel strongly…
Bias (gender, ethnic or other)	Manning levels
Humour	He talks of gearbox torque.
Unexplained acronyms	APA
Strong interruptions (em dash)	—an amazing feat—
Nominalisations	He carried out an inspection.
Slang or colloquial expressions	It was dead good.
Anthropomorphism (inanimate objects alive)	This study concludes that…
Ellipses to trail off (OK within quotation)	Once upon a time…

Table A.3 Style options example checklist

Element of style to avoid	Options
Quotation marks	Double (or single)
Highlighting words	Single quotes (or italics or bold)
Captions	Above (or below) a table, below a figure
Referencing style	APA (or MLA)
Possessive plural for words ending in 's'	s's (or s')
Oxford comma	Only for clarity (or standard)
Speech quotations	No (or yes)
(Unexplained) general comparisons	No (or yes)
Conjunctions to start a sentence	No (or yes)
Latin	No (or yes)

Table A.4 Proof-reader example checklist

Item	Checked, comment
Latest template used (or as plan)	
All pages numbered	
All figures and tables labelled	
All figures and tables referred to in text	
All references complete (no missing fields)	
All references referred to in text	
All acronyms spelt out on first use	
All acronyms defined in glossary	
All bullet points are introduced	
All bullet points consistent format and symbols	
All fonts and formatting consistent	
Company and product brand names correct	
All cross-references correct	
All links work	
Automatic tables updated (contents, figures, tables and references)	
Figures are readable	
Spelling, grammar and punctuation checked in Word	
Word accessibility check completed	
Readable writing style and presentation	
Check for homophones, misuse of prefixes	

Table A.5 Technical reviewer example checklist

Item	Checked, comment
Equations and calculations are correct	
Correct units are defined throughout	
Technical details correct (e.g. part number)	
Charts have accurate details (e.g. axes, scale)	
References are accurate and latest (or correct) issue	

Table A.6 Approver example checklist

Item	Checked, comment
Document meets the brief or plan	
Correct and latest template used	
Summary, conclusions and recommendations convey right message	
Company copyright notice	
No sensitive or proprietary information	
Security classification correct (if applicable)	
References are available to the reader	
Document ready to issue	

Table A.7 Common and engineering homophones (continued on next page)

Homophones
Bare, bear
Berth, birth
Bite, byte
Boar, bore
Board, bored
Brake, break
Buy, by
Calendar, calender
Cawed, chord, cord, cored (accord)
Ceiling, sealing
Censor, sensor
Council, counsel
Coarse, course
Crews, cruise

Table A.7 Continued

Homophones
Cymbal, symbol
Draw, drawer
Ducks, dux
Every day, everyday
Faze, phase
Forth, fourth
Grate, great
Hall, haul
Hear, here
Higher, hire
Hoes, hose
Hole, whole
ICE, ice
It, IT
Its, it's
Knew, new
Leak, leek
Mustard, mustered
Naval, navel
Non-, none
Oar, or, ore
One, wan
Pail, pale
Pair, pare, pear
Plain, plane
Read, red
Right, rite, write
Root, route
Sauce, source
Shear, sheer
Sight, site
Steal, steel
Straight, strait
Sub-, sub

(Continued)

Table A.7 Continued

Homophones
Tern, turn
Their, there, they're
To, too, two
Toe, tow
Wait, weight
Warn, worn
Weather, whether
Weir, we're
We'll, wheel
Which, witch

Table A.8 Punctuation

Punctuation mark		Use
Capital letter	A, B, C	■ First letter of a sentence ■ First letters of names of people, places, festivities and titles ■ First letters of main words in a title ■ Abbreviations or acronyms for words starting with a capital letter
Full stop (or period)	.	■ Finishing a sentence for a statement or indirect question (and is followed by a single space) ■ After abbreviations
Question mark	?	■ Direct question
Exclamation mark	!	■ Exclamation
Comma	,	■ Listing comma: separate items in a list (an Oxford comma comes before the final 'and' in a list; not generally used in British English but may be used for clarity, e.g. if there is an 'and' in the final item) ■ Joining comma: join two sentences with a conjunction ■ Bracketing comma: pair of commas (or only one at the beginning or end of a sentence) to mark a weak interruption (non-defining relative clause) in a sentence ■ Gapping comma: show a word (or words) has been omitted ■ Before or after speech
Colon	:	■ Explain or expand on something, such as introducing bullet points
Semicolon	;	■ Link two, complete, closely related sentences. No capital letter after a semicolon ■ Used to separate phrases in a list (if any item contains a comma)

Table A.8 Continued

Punctuation mark		Use
Hyphen	-	■ Split words over two lines ■ Compound words ■ With a prefix (particularly with two consecutive vowels) ■ With modifiers and multiple words
En dash	–	■ Span or range
Em dash	—	■ Strong interruption in a sentence
Apostrophe	'	■ Possession ■ Contraction
Single opening and closing quotation	' '	■ Quotation ■ 'Not-my-idea' expressions
Double opening and closing quotation	" "	■ Quotation
Brackets (or parentheses)	( )	■ Around a clause in a sentence or adding the letter 's' or explanation; always used as a pair
Square brackets	[ ]	■ To interrupt a quotation to add an explanation; always used as a pair
Ellipsis	…	■ Words omitted in a quotation or a sentence tailing off

Table A.9 Solution to rationale exercise

Example	Rationale
1. Engineering design log	record
2. Progress report	inform
3. Acceptance test specification	instruct
4. Request for proposal (RFP)	find out
5. Proposal	persuade
6. Software release note	record
7. Audit report	inform
8. Subject access request	find out
9. Troubleshooting document	guide
10. Business case	persuade
11. Onboarding documentation	guide
12. Technical service bulletins	instruct

emerald PUBLISHING ice Publishing

Jean Billingsley
ISBN 978-1-83662-927-6
https://doi.org/10.1108/978-1-83662-924-520261010

Index

Note: Italics in page numbers indicates figures and tables.

Printed and bound by CPI Group (UK) Ltd, Croydon, CR0 4YY
15/10/2026
14982700-0005